Reviews of Environmental Contamination and Toxicology

VOLUME 206

For further volumes:
http://www.springer.com/series/398

Reviews of Environmental Contamination and Toxicology

Editor
David M. Whitacre

Editorial Board
María Fernanda Cavieres, Playa Ancha, Valparaíso, Chile • Charles P. Gerba, Tucson, Arizona, USA
John Giesy, Saskatoon, Saskatchewan, Canada • O. Hutzinger, Bayreuth, Germany
James B. Knaak, Getzville, New York, USA
James T. Stevens, Winston-Salem, North Carolina, USA
Ronald S. Tjeerdema, Davis, California, USA • Pim de Voogt, Amsterdam, The Netherlands
George W. Ware, Tucson, Arizona, USA

Founding Editor
Francis A. Gunther

VOLUME 206

Coordinating Board of Editors

DR. DAVID M. WHITACRE, *Editor*
Reviews of Environmental Contamination and Toxicology

5115 Bunch Road
Summerfield, North Carolina 27358, USA
(336) 634-2131 (PHONE and FAX)
E-mail: dmwhitacre@triad.rr.com

DR. HERBERT N. NIGG, *Editor*
Bulletin of Environmental Contamination and Toxicology

University of Florida
700 Experiment Station Road
Lake Alfred, Florida 33850, USA
(863) 956-1151; FAX (941) 956-4631
E-mail: hnn@LAL.UFL.edu

DR. DANIEL R. DOERGE, *Editor*
Archives of Environmental Contamination and Toxicology

7719 12th Street
Paron, Arkansas 72122, USA
(501) 821-1147; FAX (501) 821-1146
E-mail: AECT_editor@earthlink.net

ISSN 0179-5953
ISBN 978-1-4419-6259-1   e-ISBN 978-1-4419-6260-7
DOI 10.1007/978-1-4419-6260-7
Springer New York Dordrecht Heidelberg London

Library of Congress Control Number: 2010929860

© Springer Science+Business Media, LLC 2010
All rights reserved. This work may not be translated or copied in whole or in part without the written permission of the publisher (Springer Science+Business Media, LLC, 233 Spring Street, New York, NY 10013, USA), except for brief excerpts in connection with reviews or scholarly analysis. Use in connection with any form of information storage and retrieval, electronic adaptation, computer software, or by similar or dissimilar methodology now known or hereafter developed is forbidden.
The use in this publication of trade names, trademarks, service marks, and similar terms, even if they are not identified as such, is not to be taken as an expression of opinion as to whether or not they are subject to proprietary rights.

Printed on acid-free paper

Springer is part of Springer Science+Business Media (www.springer.com)

Contents

Physiological Adaptations of Stressed Fish to Polluted Environments: Role of Heat Shock Proteins 1
Ekambaram Padmini

Phytoremediation: A Novel Approach for Utilization of Iron-ore Wastes 29
Monalisa Mohanty, Nabin Kumar Dhal, Parikshita Patra, Bisweswar Das, and Palli Sita Rama Reddy

Fugitive Dust and Human Exposure to Heavy Metals Around the Red Dog Mine 49
Elizabeth J. Kerin and Hsing K. Lin

A Profile of Ring-hydroxylating Oxygenases that Degrade Aromatic Pollutants 65
Ri-He Peng, Ai-Sheng Xiong, Yong Xue, Xiao-Yan Fu, Feng Gao, Wei Zhao, Yong-Sheng Tian, and Quan-Hong Yao

Environmental Implications of Oil Spills from Shipping Accidents ... 95
Justyna Rogowska and Jacek Namieśnik

Water Quality in South San Francisco Bay, California: Current Condition and Potential Issues for the South Bay Salt Pond Restoration Project 115
J. Letitia Grenier and Jay A. Davis

Index 149

Contributors

Bisweswar Das Institute of Minerals and Materials Technology (CSIR), Bhubaneswar 751013, Orissa, India

Jay A. Davis San Francisco Estuary Institute, 7770 Pardee Lane, Oakland, CA 94621, USA

Nabin Kumar Dhal Institute of Minerals and Materials Technology (CSIR), Bhubaneswar 751013, Orissa, India

Xiao-Yan Fu Shanghai Key Laboratory of Agricultural Genetics and Breeding, Agro-Biotechnology Research Institute, Shanghai Academy of Agricultural Sciences, 2901 Beidi Road, Shanghai, People's Republic of China

Feng Gao Shanghai Key Laboratory of Agricultural Genetics and Breeding, Agro-Biotechnology Research Institute, Shanghai Academy of Agricultural Sciences, 2901 Beidi Road, Shanghai, People's Republic of China

J. Letitia Grenier San Francisco Estuary Institute, 7770 Pardee Lane, Oakland, CA 94621, USA

Elizabeth J. Kerin University of Alaska Fairbanks, Fairbanks, AK, USA

Hsing K. Lin Mineral Industry Research Laboratory, Institute of Northern Engineering, University of Alaska Fairbanks, Fairbanks, AK, USA; Department of Resources Engineering, National Cheng Kung University, Tainan City, Taiwan

Monalisa Mohanty Institute of Minerals and Materials Technology (CSIR), Bhubaneswar 751013, Orissa, India

Jacek Namieśnik Department of Analytical Chemistry, Chemical Faculty, Gdańsk University of Technology, ul. Narutowicza 11/12, 80-233 Gdańsk, Poland

Ekambaram Padmini Department of Biochemistry, Bharathi Women's College, Chennai 600 108, Tamilnadu, India

Parikshita Patra Institute of Minerals and Materials Technology (CSIR), Bhubaneswar 751013, Orissa, India

Ri-He Peng Shanghai Key Laboratory of Agricultural Genetics and Breeding, Agro-Biotechnology Research Institute, Shanghai Academy of Agricultural Sciences, 2901 Beidi Road, Shanghai, People's Republic of China

Palli Sita Rama Reddy Institute of Minerals and Materials Technology (CSIR), Bhubaneswar 751013, Orissa, India

Justyna Rogowska Department of Analytical Chemistry, Chemical Faculty, Gdańsk University of Technology, ul. Narutowicza 11/12, 80-233 Gdańsk, Poland

Yong-Sheng Tian Shanghai Key Laboratory of Agricultural Genetics and Breeding, Agro-Biotechnology Research Institute, Shanghai Academy of Agricultural Sciences, 2901 Beidi Road, Shanghai, People's Republic of China

Ai-Sheng Xiong Shanghai Key Laboratory of Agricultural Genetics and Breeding, Agro-Biotechnology Research Institute, Shanghai Academy of Agricultural Sciences, 2901 Beidi Road, Shanghai, People's Republic of China

Yong Xue Shanghai Key Laboratory of Agricultural Genetics and Breeding, Agro-Biotechnology Research Institute, Shanghai Academy of Agricultural Sciences, 2901 Beidi Road, Shanghai, People's Republic of China

Quan-Hong Yao Shanghai Key Laboratory of Agricultural Genetics and Breeding, Agro-Biotechnology Research Institute, Shanghai Academy of Agricultural Sciences, 2901 Beidi Road, Shanghai, People's Republic of China

Wei Zhao Shanghai Key Laboratory of Agricultural Genetics and Breeding, Agro-Biotechnology Research Institute, Shanghai Academy of Agricultural Sciences, 2901 Beidi Road, Shanghai, People's Republic of China

Physiological Adaptations of Stressed Fish to Polluted Environments: Role of Heat Shock Proteins

Ekambaram Padmini

Contents

1	Introduction	1
2	Fish as Environmental Biomonitors	2
3	Impact of Environmental Stressors on Fish	3
4	Stress Biomarkers of Aquatic Contamination	4
5	Stress Response in Fish	5
6	Molecular Chaperones and HSP Families	6
7	Role of Stress Proteins in Fish	8
8	HSP Genes in Fish	11
9	Mechanistic Regulation of HSP Induction	12
10	Role of HSPs in Survival	14
11	Seasonal Influences on HSP Expression	15
12	Conclusions	16
13	Summary	17
	References	18

1 Introduction

Living systems encounter a variety of stresses at the organismal and cellular levels during their continuous interaction with the environment. Fish are particularly threatened by aquatic pollution, and the environmental stress they face may help shape their ecology, evolution, or biological systems (Sorensen and Loeschcke 2007). At the cellular level, stress can be regarded as any disturbance to normal development (Tiligada 2006). Environmental stress often activates the endogenous production of reactive oxygen species (ROS), which are an integral part of intracellular communication (Fedoroff 2006). Hence, constant exposure to stressors

E. Padmini (✉)
Department of Biochemistry, Bharathi Women's College, Chennai 600 108, TN, India
e-mail: dstpadmini@rediffmail.com

may result in ROS-mediated oxidative sequelae that irreversibly damage proteins (Eustace and Jay 2004) and thus compromise antioxidant defense, cellular function, and survival (Thomson et al. 1998). To repair such damage or eliminate damaged components, organisms have evolved the cellular stress response, which includes the induction of a highly conserved set of cytoprotective proteins called stress proteins or heat shock proteins (HSPs; Schlesinger et al. 1982).

HSPs comprise a family of highly homologous chaperone proteins that help to ensure the proper folding and formation of proteins, and maintenance of their appropriate conformations; HSPs also facilitate the intracellular localization of newly synthesized polypeptides (Cheng et al. 2006; Soti et al. 2005). HSPs interact with a wide variety of cellular proteins and thus are important components of cellular networks (Csermely 2004). In response to stress stimuli, cells produce high levels of HSPs to protect themselves against unfavorable conditions (Benjamin and McMillan 1998). The demand for molecular chaperones increases proportionately with stress status, because the rate of damage to cell proteins or problems with protein folding increases as stress increases (Sorensen et al. 2003). Padmini et al. (2008b, 2009b) reported that HSPs are rapidly synthesized by cells in response to environmental pollutant-induced oxidative stress.

Much of what is known about the biology of HSPs has been derived from a limited number of model systems. Fish are physiologically similar to mammals and typically experience long-term exposure in habitats that range from the pristine to ones that are highly polluted, making them an ideal organism for assessing the protective role of HSPs (Currie et al.1999; Sherry 2003). HSP expression has been examined in many species (Feder and Hofmann 1999), including fish (Iwama et al. 1998), and the majority of such studies have been performed in the laboratory. Furthermore, those studies that have addressed the impact of environmental stressors or contaminants on oxidative stress markers in fish in vitro did so by exposing the animals to a single anthropogenic stress under controlled conditions, neglecting the cumulative effect of other stressors to which an organism is actually subjected under natural conditions. In addition, knowledge about seasonal variation in HSP induction remains negligible. In this chapter, I endeavor to address the impact of different types of environmental stressors on fish and provide an update on the role of HSPs in the adaptive response of stressed fish, with a focus on associated seasonal effects.

2 Fish as Environmental Biomonitors

An aquatic environment is characterized by marked temporal and spatial heterogeneity in oxygen content. Oxygen content is affected by water temperature, salinity, degree of mixing of surface and deep water masses, vertical stratification of the water column, eutrophication, and the intensity of respiration rates of aquatic organisms. Simultaneously, a staggering array of anthropogenic chemicals exists in many aquatic ecosystems, whose inhabitants then absorb and concentrate these contaminants in their tissues. Consequently, such organisms constitute the main pathway of environmental contaminant export from the aquatic to terrestrial systems (via the food chain), ultimately reaching human beings (typically in concentrated form via biomagnification). Fish are an important component of aquatic food chains as well as the human diet and thus are a suitable and practical group of animals for assessing the pollution status of aquatic ecosystems through biomonitoring activities (Padmini et al. 2004).

3 Impact of Environmental Stressors on Fish

Fish are exposed to stressors in the natural environment, in aquaculture, and intentionally under artificial conditions in the laboratory. Iwama et al. (2004) categorized stressors into three groups that greatly impact ectotherms, namely, environmental (e.g., pollutants), physical (e.g., handling, transport), and biological (e.g., disease). The present review focuses on the first (environmental) category. Niu et al. (2008), Poltronieri et al. (2007), and Vijayan et al. (1997) provide details on the impact of physical stressors on fish, and Cappo et al. (1995) discusses an important example of the effects of certain biological stressors. Environmental stressors may include natural, if sometimes extreme conditions and variable water quality parameters such as dissolved oxygen, ammonia, hardness, pH, gas content, and partial pressures (Iwama et al. 2004; Padmini and Usha Rani 2009b). For example, intertidal species are good models for studying the effects of short-term changes in the thermal history on cellular stress response, because they are exposed to daily (or hourly) fluctuations in water quality and temperature (Basu et al. 2002). Similarly, endemic Antarctic fish with cold-adapted physiologies are stenothermic (i.e., can tolerate only a very narrow range of temperatures); their upper lethal temperatures are 8–12°C, making them an ideal system for studying the biochemical and molecular mechanisms of adaptation to low temperatures (Buckley et al. 2004). Padmini and Vijaya Geetha (2007a) demonstrated the stressed status of fish in estuaries having different physicochemical parameters. However, one of the most threatening and well-documented environmental stressors is environmental pollutants. For example, contaminants such as polycyclic aromatic hydrocarbons, polychlorinated biphenyls, arsenic, chlorine, cyanide, and various phenols are potent stressors for all salmonid species. Heavy metals such as chromium, manganese, iron, nickel, copper, zinc, selenium, lead, cadmium, and mercury, the leading environmental metallic pollutants, exert significant stress impact on estuarine gray mullet *Mugil cephalus* (Padmini and Usha Rani 2009b; Padmini and Vijaya Geetha 2007b). High concentrations of metals such as copper, cadmium, zinc, and iron may also cause death in exposed fish. Other potential environmental contaminants include insecticides, herbicides, fungicides, and defoliants. Increasing human population and urbanization, as well as agricultural and industrial activities, contribute contaminants to the environment that can affect fish at all life stages (Dhaliwal and Kukal 2005).

4 Stress Biomarkers of Aquatic Contamination

Oxidative stress is a state of unbalanced tissue oxidation characterized by a disturbance in ROS production and antioxidant systems (Abele and Puntarulo 2004). ROS are oxygen free radicals that include partially reduced intermediates produced during the reduction of oxygen to water; these include superoxide anions ($O_2^{\cdot-}$), hydroxyl radicals ($^{\cdot}OH$), and non-radical active species such as hydrogen peroxide (H_2O_2; Abele and Puntarulo 2004). These free radicals are highly reactive (Buettner and Schafer 2000), because they harbor an unpaired reactive electron that can produce toxic effects. The redox cycling of heavy metals and their interactions with organic pollutants are major contributors to the oxidative stress that results from pollution in aquatic environments (Padmini et al. 2009a).

Research results have demonstrated that high concentrations of metals in water can be correlated with the stress status of fish via the redox properties of these metals (Padmini and Vijaya Geetha 2007a, 2008). Fish absorb heavy metals via their gills, the main site of xenobiotic transfer, and these toxins spread throughout the body, potentially causing deleterious effects on target organs (Buet et al. 2002). Oxidative stress renders significant damage to various biochemical processes. For example, our previous studies on oxidative stress in *M. cephalus,* which inhabits polluted environments, demonstrated significant effects on erythrocytes (Padmini et al. 2006) as well as various organs including the gill (Padmini and Sudha 2004), brain (Padmini and Kavitha 2005a, b, 2007), and liver (Padmini et al. 2004). The impact of pollution-induced oxidative stress has also been assessed by genotoxicity studies (Padmini and Kavitha 2005b). The accumulation of damaged and oxidized macromolecules such as lipids, proteins, and DNA, in various organs during oxidative stress, may decrease reproductive success, increase susceptibility to infection, and lead to the sudden death of fish in large numbers (Padmini and Sudha 2004; Padmini et al. 2004). Indeed, high levels of oxidized cellular macromolecules are correlated with shorter cell life spans (Barja and Herrero 2000). Consistent with the latter report, elevated oxidative and nitrative stresses, and their association with decreased viability of hepatocytes from fish in polluted estuaries, have been well documented (Padmini and Vijaya Geetha 2009a; Padmini et al. 2008b).

The induction of antioxidant enzymes is an important line of defense against free radical-induced oxidative stress in biological systems (Parihar et al. 1997). Antioxidant systems such as superoxide dismutase (SOD), catalase (CAT), glutathione peroxidase (GPx), glutathione reductase (GRd), and glutathione *S*-transferase (GST) protect against chemical reactive species (CRS) produced by endogenous metabolism or the biotransformation of xenobiotics. These systems may be induced, enhanced, or inhibited in response to chemical stress. Such induction can be considered an adaptation that allows an organism to partially or totally overcome environmental stress. In contrast, a deficiency of antioxidants may result in a precarious state, in which toxicity is enhanced, and a species becomes susceptible to toxic action. ROS-detoxifying enzymes may help protect against oxidative stress, and thus extend the life span of animal cells (Keaney et al. 2004; Parker et al. 2004). For example, increasing the activities of cytosolic CuZnSOD and CAT via transgenic overexpression increases cell viability by up to 33% in *Drosophila melanogaster* (Orr and Sohal 1994; Schriner et al. 2005). GRd activity in both gill and *head kidney* tissue of fish was reported to increase with increasing site contamination (Dautremepuits et al. 2009). Hence, antioxidant enzymes are thought to provide enhanced resistance to oxidative stress.

One way in which fish are used as environmental monitors is by tracking levels of their antioxidant enzymes as biomarkers of aquatic contamination (Eufemia et al. 1997; Fenet et al. 1996; Rodriguez-Ariza et al. 1993; Van der Oost et al. 1996). Antioxidant systems have long been studied in fish and bivalves exposed to chemicals experimentally or collected from polluted areas (Lenaire and Livingstone 1993; Stegeman et al. 1992). Researchers studying the freshwater bivalve *Unio tumidus* used multiple antioxidant parameter measures as indices of disturbance that resulted from exposure to contaminated sediments (Cossu et al. 1997; Doyotte et al. 1997). The role of antioxidants as useful biomarkers of pollution, both under laboratory and field conditions, has been demonstrated in freshwater fish species such as *Channa punctatus* Bloch., *Heteropneustes fossilis* (Bloch), and *Wallago attu* (Bl. and Schn) (Ahmad et al. 2000; Pandey et al. 2001, 2003). In these studies, changes in the activities of several different antioxidant enzymes (SOD, CAT, GPx, GRd, and GST) were compared to the degree of contamination of the aquatic ecosystems in which these species lived. Wilhelm Filho et al. (2001) studied the influence of pollution on the antioxidant defenses of the cichlid *Geophagus brasiliensis* and suggested that the use of biochemical indicators in environmental pollution studies is of high toxicological relevance and oxidant-mediated responses are useful indices of environmental quality. There are also cases of a negative correlation having been established between oxidative stress and both antioxidant defense status and the activities of detoxification enzymes, in polluted environments (Padmini and Vijaya Geetha 2007c; Padmini et al. 2008a, 2009a). Lipid peroxidation and subsequently reduced antioxidant defense, noted under cytotoxic conditions, support the use of such parameters as biomarkers of toxicity (Cossu et al. 2000; Telli Karakoc et al. 1997).

5 Stress Response in Fish

In addition to the induction of antioxidant enzymes as a response to chemical stressors, fish also exhibit a generalized physiological stress response. This response, which involves the activation of the hypothalamic–pituitary–interrenal axis (HPI), is common to environmental, physical, and biological stressors and helps to maintain the animal's normal or homeostatic state (Barton 2002; Iwama et al. 1999). This generalized stress response is considered to be adaptive and to represent the natural capacity of the fish to respond to stress. The physiological responses of fish to stressors have been broadly categorized as primary, secondary, and tertiary. The initial or primary response represents the perception of an altered state and initiates an endocrine response that forms part of the generalized stress response in fish (Gamperl et al. 1994). This includes the rapid release of stress hormones, such as catecholamines and cortisol from the chromaffin tissue into the circulation; the physiological action of these hormones depends on the presence of appropriate receptors on target tissues. The secondary response comprises a wide range of mechanisms involving stress hormones (Vijayan et al. 1994) in the blood, organs, and tissues of the animal (Barton et al. 2002). The tertiary response represents the whole-animal and population-level changes associated with stress. For example, when a fish is unable to adapt to stress, whole-animal changes may occur as a result of energy repartitioning that allows the animal to cope with stress-induced increased energy demand (Iwama et al. 2004). Hence, chronic exposure to stressors, depending on intensity and duration, can lead to decreases in growth, disease resistance, reproductive success, swimming performance, and other characteristics of the animal. In addition to these whole-animal changes, fish also respond to stressors at the cellular level such as the one characterized by changes in HSP concentrations.

6 Molecular Chaperones and HSP Families

Molecular chaperones are major cell constituents that exist in all organisms under non-stress conditions. They are essential to protect certain proteins against aggregation, and they help in the folding of nascent proteins or refolding of damaged proteins. They also dissolve loose protein aggregates, sequester excessively damaged proteins to larger aggregates, and target severely damaged proteins for degradation. The need for molecular chaperones increases under stressful conditions (Sorensen et al. 2003), during which time they utilize a cycle of ATP-driven conformational changes to fold or refold their targets (Papp et al. 2003). HSPs maintain the proper function of proteins and constitute a subset of molecular chaperones. They are a super family of highly conserved, ubiquitous, intracellular proteins that are responsible for diverse cellular processes such as protein folding, activation, transport, and oligomeric assembly (Csermely et al. 1998). HSPs protect against environmental and physiologic stress as well as the deleterious consequences of an imbalance in protein homeostasis, as many stresses, if prolonged, result in defective development and pathologies associated with a diverse array of diseases (Morimoto and Santoro 1998).

A stressor can be any sudden change in the cellular environment to which the cell is not prepared to respond. Almost all types of cellular stressors induce HSPs (Beere 2004; Soti et al. 2005), and several different types of environmental stressors may trigger HSP overproduction; these include infection, inflammation, exercise, exposure to toxins (e.g., ethanol, arsenic, trace metals, and ultraviolet light, among many others), starvation, hypoxia, and water deprivation. HSP induction in response to these different stressors has been confirmed by in vivo and in vitro studies, and the unique nature of HSP synthesis under such situations is correlated with the acquisition of thermotolerance and cytoprotection (Sreedhar and Csermely 2004).

HSP induction can be triggered by sudden increases in the amount of damaged proteins or abnormal polypeptides in the cytosol or nucleus (by the inhibition of their elimination via the proteasome) as well as by damage to the chaperones themselves.

HSPs are encoded by multigene families, range in molecular size from 10 to 150 kDa, and are found in all major cellular compartments. They are divided into the following families, in which the number represents the lowest molecular weight range of its members: HSP10, small HSPs (~15–35 kDa), HSP40, HSP60, HSP70, HSP90, and HSP110. Each family includes one or more members; HSP10, HSP60, and HSP75 are mainly located in mitochondria, whereas others are present in the cytoplasm and nucleus (Hartl and Hayer-Hartl 2002). Although the primary physiological function of all HSPs is to fulfill full chaperoning activity, each HSP plays a specific role. For example, HSP10 is an essential component of the mitochondrial protein-folding apparatus and participates in various aspects of HSP60 function. Some small HSPs, such as HSP22 and HSP23, which exist in oligomeric complexes of 200–800 kDa, serve as molecular chaperones that promote F-actin assembly. HSP27, another member of a small HSP family, plays a key role in assisting the assembly of macroglobular protein complexes, such as F-actin polymerization. HSP40 helps to bind and transport collagen. Its isoforms are primarily involved in the regulation of HSP70 chaperone activity; in association with either heat shock cognate (HSC)70 or HSP70, HSP40 protects macrophages from NO-mediated apoptosis (Gotoh et al. 2001). HSP60 forms a large hetero-oligomeric protein complex called the TCP1 ring complex, which is essential for assembling polypeptides, translocating proteins across membranes, and accelerating protein folding and unfolding. HSP70 is one of the most widely examined families, whose members differ in their spatial and subcellular distribution (Rohde et al. 2005). In addition to its chaperone functions in response to stressful conditions such as heat shock, hyperosmotic stress, oxidative stress, UV radiation, amino acid analogues, infection, inhibitors of energy metabolism, and heavy metals (Morimoto et al. 1992), HSP70 protects cells from a number of apoptotic stimuli (Li et al. 2000). HSP90, which constitutes 1–2% of the total protein in the cell, is ubiquitously expressed under normal physiological conditions and is predominantly localized in the cytoplasm (Nollen and Morimoto 2002; Whitesell and Lindquist 2005). There are two major isoforms, HSP90α and HSP90β, which share 78% homology (Sreedhar et al. 2004; Whitesell and Lindquist 2005). HSP90 contributes to the folding of various cellular proteins and modulates the activity of a vast number of client proteins involved in cell survival and death pathways (Buchner 1999; Picard 2002; Sreedhar et al. 2004). It is active in supporting various components of the cytoskeleton and steroid hormone receptors (Young et al. 2001) and has wide substrate specificity that includes transcription factors, kinases, and polymerases (Wegele et al. 2004). HSP110 is a cytosolic HSP, and a distant relative of the HSP70 family that cooperates with HSP70 in protein folding in the eukaryotic cytosol, catalyzing efficient nucleotide exchange on HSP70 (Dragovic et al. 2006; Raviol et al. 2006). Hence, the involvement of HSPs in various aspects of protein metabolism (Fink and Goto 1998) is essential for regulating cellular homeostasis and promoting cell survival (Bukau and Horwich 1998; Hartl 1996). HSP-induced cytoprotection can also be partly attributed to the suppression of apoptosis (Samali and Orrenius 1998).

7 Role of Stress Proteins in Fish

Highly expressed HSPs have been documented in all cells that have been examined, from bacteria to mammals (Csermely et al. 1998; Feder and Hofmann 1999), including fish (Iwama et al. 1998). HSPs in fish cells, similar to those in other vertebrates, are upregulated in response to different types of stressors. Delaney and Klesius (2004) reported the induction of HSP70 in the blood, brain, and head-kidney of juvenile Nile tilapia (*Oreochromis niloticus* (L.)) during exposure to hypoxic conditions. Many studies have documented HSP induction in response to aquatic contaminants (Ryan and Hightower 1996; Vijayan et al. 1998; Williams et al. 1996). β-Naphthoflavone (BNF), a potent inducer of P450 enzymes, induces HSP70 expression in the liver tissue of rainbow trout, *Oncorhynchus mykiss* (Vijayan et al. 1997). Sublethal concentrations of bleached kraft pulp mill effluent (BKME) and sodium dodecyl sulfate (SDS) result in higher hepatic HSP70 expression in the salmonid liver, suggesting a chronic cellular stress response associated with contaminant exposure (Vijayan et al. 1998). HSP induction in response to hyperammonemic conditions has been documented in carp (Hernandez et al. 1999). Enhanced levels of different HSPs have also been demonstrated in tissues of fish exposed to polycyclic aromatic hydrocarbons (Vijayan et al. 1998), metals such as copper, zinc, and mercury (Duffy et al. 1999; Williams et al. 1996), pesticides (Hassanein et al. 1999), and arsenite (Grosvik and Goksoyr 1996). Cara et al. (2005) reported the impact of food-deprivation, reduced oxygen levels, and heat shock on the expression of HSP70 and HSP90 in the early life stages of the gilthead sea bream (*Sparus aurata*), a warm-water aquaculture species. Clarkson et al. (2005) reported that HSP90 is important after exhaustive exercise, in a manner independent of its role in protection against cellular damage. Hence, the induction of different types of HSPs varies depending on the type of tissue and type and nature of stressor (Airaksinen et al. 2003; Rabergh et al. 2000; Smith et al. 1999). In addition, the sensitivity of HSP expression may vary among species (Basu et al. 2001; Nakano and Iwama 2002) and developmental stage (Martin et al. 2001; Santacruz et al. 1997).

Basu et al. (2002), in a previous review, have demonstrated that HSPs play a key role in various aspects of fish physiology, including development, aging, stress physiology, endocrinology, immunology, environmental physiology, adaptation, and stress tolerance. Researchers have reported the induction of various HSP families in fish following exposure to various environmental stressors and have demonstrated that these HSPs provide a protective role (Currie et al.1999; Sherry 2003). Other studies have affirmed the use of the HSP response as an indicator of stress status in fish (Samali and Orrenius 1998; Sreedhar and Csermely 2004). HSPs have a relatively short half-life (6–9 hr in *Drosophila*; Lindquist 1986), but their levels remain elevated in organisms long after the stressor is removed. Hence, HSPs may play a role in the long-term adaptation of animals to their environment (Morimoto and Santoro 1998). Indeed, a positive and direct correlation between HSP expression and thermotolerance has been well documented (Russotti et al. 1996). Mosser and Bols (1988) also showed that the appearance and decay of HSPs share a close temporal relationship with the induction and disappearance of thermotolerance.

The results of many in vitro studies have demonstrated the induction of HSPs and their role in stress resistance, using cell lines (Kothary et al. 1984; Mosser and Bols 1988), primary cell cultures (Sathiyaa et al. 2001), and whole animal tissues (Ackerman et al. 2000; Dietz and Somero 1993). Ryan and Hightower (1994) demonstrated metal-induced dose-dependent changes in stress protein levels in response to environmental stressors using a hepatoma-derived cell line (PLHC-1) from desert topminnow (*Poeciliopsis lucida*) and primary winter flounder (*Pleuronectes americanus*) kidney cultures. Kong et al. (1996) examined the expression patterns of different HSPs such as HSP70, HSP40, and HSP45, in Chinook salmon embryonic cells (CHSE-214), by exposing them to heat treatment at 24°C at time periods ranging from a few minutes to 24 hr. Martin et al. (1998) found that high expression levels of HSP60 and HSP70, in the neural tissue of four different teleostean fish species exposed to acid shock for 2 hr at pH 4.5, were correlated with increased acidosis resistance. Taken together, these results demonstrate a correlation between enhanced levels of HSPs and the extent of exposure to stressors within an ecologically relevant range. These reports also confirm the importance of HSP expression for the maintenance of cellular homeostasis (Cheng et al. 2006) and have helped elucidate their role in enhancing the health and survival of stressed fish (Basu et al. 2002).

Most studies that address HSPs and cellular stress response do so under laboratory conditions by subjecting cells or organisms to a known concentration of one or more stressors. However, fish are challenged with multiple types of environmental stressors in their natural environment, and only a few studies have assessed the impact of an aggregate of environmental stressors on aquatic organisms. Such studies on fish HSP expression under natural conditions are very important in elucidating the functional significance of these proteins to environmental stress tolerance and species adaptation and are essential if fish are to be used in aquatic ecosystem biomonitoring. The small stress protein, HSP30, has been shown to react to protein-damaging stressors in fish tissues, most notably to heat shock, which leads to significant increases in mRNA and protein levels (Currie et al. 2000; Lund et al. 2002). A significant increase in inducible HSP70 expression serves as a sensitive indicator of the cellular stress response associated with exposure to contaminants. Induction of HSP70 in BNF-exposed trout correlates with altered hepatic function, suggesting that HSP70 expression may be used as an indicator of the cellular effects of the toxicant (Vijayan et al. 1997). Higher HSP70 expression, after exposure to sublethal concentrations of toxic BKME and SDS, may help cells/tissues cope with the toxic insult (Vijayan et al. 1998). In the same study, the absence of a significant increase in HSP70 expression in cells/tissues exposed to 32 mg/L SDS, which resulted in mortality, suggested that HSP70 induction may be closely correlated with the ability to protect cellular function and consequently animal survival. Hence, these authors demonstrated that HSP70 expression is a sensitive indicator of the cellular stress response associated with sublethal concentrations of contaminants, and that its expression in fish tissue may be a useful biomarker of aquatic pollution. Padmini and Usha Rani (2008) reported that enhanced levels of HSP70 in grey mullet (*M. cephalus*) from polluted sites may reflect a protective response against environmental pollutant-related stress and suggested that higher levels of HSP70 may be an adaptive strategy to maintain native protein structures under environmental stress. Iwama et al. (2004) addressed the functional importance of HSP70 in thermal tolerance using two intertidal fishes as model species, the tide-pool sculpin (*Oligocottus maculosus*) and fluffy sculpin (*Oligocottus snyderi*). They found that the levels of constitutive HSP70 and the associated scope for increase in HSP correlated with the ability of tidepool sculpins to cope with environmental changes. It was recently shown in *M. cephalus* that significant induction of HSP70 and HSP90 in liver mitochondria and hepatocytes, respectively, plays an important role in tolerance against pollution-induced oxidative stress (Padmini et al. 2008b, 2009b). The role of HSPs in thermotolerance appears to be crucial, because the inhibition of their synthesis prevents the development of thermotolerance in rainbow trout (*O. mykiss*) fibroblasts (Mosser and Bols 1988). Studies employing transgenic *Drosophila* that overexpress HSP70 support the idea that HSPs are at least partially responsible for induced thermotolerance (Feder and Krebs 1998). Basu et al. (2002) also demonstrated that HSPs contribute significantly to thermal adaptation through genetic mechanisms. Although the cellular stress response is generally accepted as a common feature of a wide variety of organisms (Feder and Hofmann 1999), some animals, under extreme conditions, lose the ability to invoke a measurable cellular stress response. HSP70 levels in the brain, white muscle, gill tissues (Carpenter and Hofmann 2000), and hepatocytes (Hofmann et al. 2000) of the Antarctic fish *Trematomus bernacchii* do not significantly increase when fish are subjected to a thermal challenge. Because the thermal environment of the Antarctic Ocean is extremely stable (Carratu et al. 1998) and very little chemical pollution exists in this region, Hofmann et al. (2000) concluded that there might not have been strong selective pressure for these fish to retain a potentially energetically costly stress response. Therefore, these authors suggest that Antarctic fish may have lost the ability to respond to heat stress at a genetic level, and that changes in environmental temperature may not have a significant effect on the cellular stress response and thermal tolerance of *T. bernacchii*.

HSPs appear to have many developmental roles in addition to thermotolerance and cytoprotection. For example, they have specific cellular functions during embryogenesis that serve to sustain the development of the whole organism rather than solely maintaining cellular processes for "housekeeping" (Morange 1997). Krone and Sass (1994) observed low levels of constitutive HSP90α in developing zebrafish, but the gene which encodes this protein was strongly induced following heat stress in gastrula- and late-stage embryos. However, constitutive levels of HSP90β were high relative to HSP90α, but only weakly induced following similar heat stress. Hence, within a specific HSP family such as HSP90, isoforms may be differentially regulated and, as a consequence, may serve different cellular functions during developmental and later stages. Localization studies in zebrafish have also revealed that constitutive HSP90α is restricted primarily to regions involved in myogenesis, including the pectoral fin buds and a subset of cells within the somites (Sass et al. 1996). HSP47 is highly induced in response to stress (Pearson et al. 1996), similar to HSP90α, and this may aid in the formation of fish embryonic tissues from its association with procollagen (Krone et al. 1997; Lele et al. 1997). Lele et al. (1997) reported that basal levels of a stress-inducible HSP70 were low during embryogenesis, but were significantly elevated in a tissue- and stress-dependent manner, when zebrafish were exposed to various stressors. Santacruz et al. (1997) also reported a constitutive and stress-induced form of HSP70 in developing zebrafish. Together, these studies suggest that HSPs are differentially expressed in a spatial, temporal, and stress-specific manner and probably serve specific roles in embryonic development.

Cross-protection, also known as cross-tolerance, is the ability of one stressor to transiently increase the resistance of an organism to a subsequent heterologous stressor. This may be a critical feature of the cellular stress response in an environmental context. Early studies of cross-protection in fish were performed in vitro in the winter flounder, *P. americanus* (Basu et al. 2002). Exposure of the renal epithelium to heat stress protected these cells against the deleterious effects of a subsequent extreme temperature or chemical challenge, and the protection afforded by a mild heat shock coincided with increased levels of HSP28, HSP70, and HSP90 in the cell (Brown et al. 1992). Dubeau et al. (1998) reported that heat-stressed Atlantic salmon (*Salmo salar*) with artificially elevated levels of brachial and hepatic HSP70 were better able to tolerate an osmotic challenge, relative to control fish, suggesting a role for HSP in ionic and osmotic adaptation. These studies provide strong evidence of the ability of one stressor to condition a fish to better tolerate a subsequent, more severe stressor. Moreover, these studies demonstrate that fish regulate HSP gene expression to enhance their tolerance to an upcoming environmental change.

8 HSP Genes in Fish

Although the roles of HSPs in fish at the protein level have been investigated in many studies, very little is known about the genes that encode HSPs in fish (Basu et al. 2002). HSP/HSC genes have been cloned or sequenced and characterized only from a modest number of fish species, primarily from families with low molecular weight. These include the hsp30 gene, cloned from Chinook salmon (*Oncorhynchus tshawytscha*; Hargis et al., unpublished data; accession number U19370); hsp47, cloned and characterized from zebrafish (*Danio rerio*; Pearson et al. 1996); hsp27 and hsp30, cloned from the desert pupfish (*P. lucida*; Norris et al. 1997); hsp70, cloned from rainbow trout (*O. mykiss*; Kothary et al. 1984), medaka (*Oryzias latipes*; Arai et al. 1995), zebrafish (Lele et al. 1997), tilapia (*Oreochromis mossambicus*; Molina et al. 2000), and pufferfish (*Fugu rubripes*; Lim and Brenner 1999); hsc71, a constitutively expressed cognate member of the hsp70 multigene family, isolated and characterized from rainbow trout (Zafarullah et al. 1992); hsc70, cloned and characterized from zebrafish (Santacruz et al. 1997) and sequenced from carp (*Cyprinus carpio*; Yin et al. 1999); and hsp90 (both alpha and beta), sequenced from zebrafish (Krone and Sass 1994) and sequenced (hsp90α only) from Chinook salmon (Palmisano et al. 2000) and Japanese flounder (*Paralichthys olivaceus*; Nam et al., unpublished data; accession number AU090921). Pan et al. (2000) characterized an hsp90 sequence from Atlantic salmon (*S. salar*) and correlated their results to the hsp90β of zebrafish, which showed 92% amino acid identity.

In recent years, many researchers have analyzed transcriptomic responses to model toxicants using novel genomic technologies to develop new versatile diagnostic tools to identify adverse biological effects of environmental pollutants on fish in coastal and marine environments. For example, Falciani et al. (2008) developed hepatic transcriptomic profiles of European flounder (*Platichthys flesus*) inhabiting different polluted sites, and, using gene expression fingerprints obtained by bioinformatics and computational approaches, correlated the results to variation in responses to chemical pollutants, indicating the potential utility of this method in assessing environmental impacts. Similarly, Williams et al. (2008) demonstrated the utility of using the microarray technique to identify toxicant-responsive genes and discriminate between modes of toxicant action using the transcriptomic responses of European flounder to model toxicants. Williams et al. (2006, 2007), during the development of GENIPOL (genomic tools for biomonitoring of pollutant coastal impact), described the establishment of a useful resource for ecotoxicogenomics and the identification of the temporal molecular responses of European flounder to cadmium and β-estradiol. Sheader et al. (2006) used custom cDNA microarray analysis of the European flounder liver to demonstrate an oxidative stress response to cadmium, and they highlighted the potential use of candidate genes as novel biomarkers, suggesting the applicability of a custom microarray approach to study the effects of toxicants. Transcriptome analysis using cDNA library clones from the brain mRNA of channel catfish (*Ictalurus punctatus*) revealed different types of genes, including stress-induced genes (Ju et al. 2000). Through integrated transcriptomic and gene expression analysis, Chen et al. (2008) identified the cellular and genomic mechanisms involved in the adaptation of Antarctic notothenioid fish to cold in the frigid southern ocean. They also demonstrated that about 177 gene families were specifically augmented in these fish, and that the genes were involved in an array of biological processes including protein synthesis, protein folding and degradation, antioxidation, and antiapoptosis, representing a collective stress-responding or stress-mitigating phenotype to overcome various physiological changes posed by freezing and an oxygen-rich environment.

9 Mechanistic Regulation of HSP Induction

The heat shock response is primarily regulated at the transcriptional level and is mediated by a family of heat shock factors (HSFs) that interact with a specific regulatory element, the palindromic (CNNGAANNTTCNNG) heat shock element

(HSE) present in the promoters of HSP genes (Pirkkala et al. 2001). At present, four different HSFs have been identified: HSF1, HSF2, HSF3, and HSF4, which are products of the transcription of four different genes (Snoeckx et al. 2001). The regulation of HSF1 activity is complex and is determined, in part, by intrinsic properties of the molecule, notably by stress-induced changes in its conformation (Farkas et al. 1998; Tomanek and Somero 2002). HSPs may keep HSF1 "locked" in the multiprotein complex under non-stressful conditions and thereby prevent it from binding to the HSE and stimulating the transcription of HSP genes; upon environmental or physiological stresses that cause proteins to unfold, this complex dissociates, because HSPs begin to bind preferentially to unfolded proteins rather than to HSF1 (Parsell and Lindquist 1993, 1994). The activation of HSF1 is characterized by the conversion of this factor from a monomeric to trimeric state (Fernandes et al. 1994). HSF1 trimers are able to move to the nucleus, where they are hyperphosphorylated and then bind to the HSE, inducing the synthesis of HSPs. Both mechanisms that promote nonnative proteins, and are responsible for the signal transduction events resulting in HSF1 trimerization, may be related to the disruption of thiol homeostasis by stressors, which thereby promotes intracellular oxidative damage (Freeman et al. 1990). Currie and Tufts (1997) first suggested that HSP70 induction in rainbow trout may be regulated primarily at the transcriptional level. An HSF1-like factor was also demonstrated to be involved in the induction of HSP70 mRNA in rainbow trout (Airaksinen et al. 1998). Rabergh et al. (2000) cloned these transcription factors in zebrafish and bluegill sunfish (*Lepomis macrochirus*). Two forms of HSF1 transcript were detected using reverse transcription-polymerase chain reaction (RT-PCR) in zebrafish. The expression of both transcripts was confirmed by RT-PCR analysis of control and heat-shocked hepatic, gonad, and gill tissues, and there were differences in the amounts of these two transcripts among tissues and in their responses to heat stress (Rabergh et al. 2000).

The effects of stress on HSF1 activation have been evaluated in several studies. One study, in which the authors carefully analyzed the effects induced by 13 different stress-response inducers including heat shock (Zou et al. 1998), showed that all agents and conditions triggered the oxidation of thiol-containing molecules, particularly glutathione. These effects led to the formation of glutathione disulfide (GSSG), mixed glutathione–protein disulfides, and protein–protein disulfides, which thereby induced trimerization of HSF and promoted their DNA-binding ability. The early transduction pathway that leads to HSF1 activation appears to involve two steps: (i) disruption of intracellular thiol-disulfide redox homeostasis, which results in the formation of disulfide-linked aggregates of cellular proteins, and (ii) recognition of denatured proteins by preexisting protein chaperones, a phenomenon that triggers HSF1 trimerization (Zou et al. 1998). Padmini and Usha Rani (2009a) showed that, in a prooxidative environment characterized by glutathione depletion, HSF1 undergoes trimerization and activation that induces HSP90α synthesis. This study suggests a critical role for the altered glutathione redox pair in HSP induction through the redox-dependent activation of HSF1. In addition, the response of HSP90 to geldanamycin (GA) exposure appears to be indirectly mediated by the accumulation of incorrectly folded proteins and/or the activation of

HSF1 (Sathiyaa and Vijayan 2003). Upregulation of HSF1 during overexpression of HSPs under stressed conditions suggests that this nuclear factor will be a useful tool to monitor natural stress.

10 Role of HSPs in Survival

In addition to the suppression of protein damage, HSPs significantly contribute to protective functions as important mediators of intracellular signaling. Under unfavorable conditions, stress kills cells not only by irreversible damage to critical structures of cell proteins, but also by activating a programmed cascade of cell death events. HSPs protect cells from such situations by interfering with programmed cell death (Gabai and Sherman 2002). HSPs play a key role in cell survival by providing an adaptive mechanism in response to different types of stress. Numerous authors have attributed the survival-promoting effects of HSPs to their ability to suppress apoptosis in response to several stimuli, including heat, DNA damage, and environmental stress (Beere 2004).

HSP27 is a novel and important factor that protects against oxidative stimuli by blocking an important initiation factor of apoptosis (Arrigo 2001). HSP27 associates with Akt and protects its kinase activity from heat stress and serum deprivation in PC12 embryonic carcinoma cells (Mearow et al. 2002). HSP27 binds to cytochrome *c* (released by mitochondria) in the cytosol and prevents the cytochrome *c*-mediated interaction of Apaf-1 with procaspase-9, thereby interfering specifically with the mitochondrial pathway of caspase-dependent cell death (Bruey et al. 2000). HSP27 inhibits the release of the proapoptotic molecule Smac/DIABLO (Chauhan et al. 2003). It also inhibits cytotoxicity induced by tumor necrosis factor (TNF)-α and inflammatory cytokines via ROS production, thereby maintaining mitochondrial integrity, supporting its role in protecting mitochondria during the activation of apoptosis (Samali et al. 2001). Xanthoudakis et al. (1999) demonstrated a role for HSP60 and HSP10 in regulating apoptosis in the presence of cytochrome *c* and dATP, in cell-free systems. It has also been shown that antisense oligonucleotide-induced decreases in mitochondrial HSP60 induce the release of cytochrome *c* and precipitate apoptosis (Knowlton and Gupta 2003; Shan et al. 2003). Modulation of ASK 1 (apoptosis signal-regulating kinase1) expression during overexpression of HSP70 appears to be a prosurvival mechanism in stressed fish liver mitochondria (Padmini and Vijaya Geetha 2009b).The co-chaperone CHIP (carboxyl terminus of Hsc70-interacting protein) is known to play an important role in this process (Demand et al. 2001). Sreedhar and Csermely (2004) also reported that HSP72 is a direct inhibitor of ASK1, and that the accumulation of HSP72 is necessary for c-jun NH_2 terminal kinase (JNK) 1/2 downregulation, suggesting that the chaperone activity of HSP70 is required for the inhibition of apoptosis. HSP70 overexpression also reduces Fas-induced apoptosis (Schett et al. 1999). Similar to HSP27, HSP70 exhibits an inhibitory role against cytochrome *c*/dATP-mediated caspase activation. It suppresses apoptosis by directly associating with Apaf-1 and blocking the assembly of a functional apoptosome (Beere et al. 2000). HSP70 also attenuates nitric oxide (NO)-induced apoptosis in RAW264.7 macrophages, thereby maintaining mitochondrial integrity via the upregulation of intracellular glutathione (Calabrese et al. 2002; Sreedhar et al. 2002). Zhang et al. (2005) reported that HSP90, together with its client protein Akt, function to inhibit ASK1-p38 signaling. HSP90 is an essential component for the activation of Akt kinase, which activates telomerase and thus acts against apoptosis (Haendeler et al. 2003). HSP90α is hepatoprotective, favoring its survival during oxidative stress by regulating ASK1 expression and thereby functionally antagonizing the cell death-promoting functions of JNK1/2 in natural aquatic systems (Padmini and Usha Rani 2010). Raf-1 forms a complex with HSP90, and inhibition of this multimolecular complex leads to the destabilization of Raf-1, thereby blocking the Raf-1–MEK–MAPK (mitogen-activated protein kinase (MAPK)/extracellular signal-regulated kinase (ERK) kinase) signaling pathway (Schulte et al. 1996). Dissociation of the HSP90–Raf-1 complex is also reported to result in apoptosis in mast cells (Cissel and Beaven 2000) and in B-lymphocytes (Piatelli et al. 2002). Moreover, HSP90 directly binds to Apaf-1 to prevent the formation of the apoptosome complex (Pandey et al. 2000). The reactive cysteines present on HSP90 are able to reduce cytochrome c, suggesting a role for HSP90 in modulating the redox status in resting and apoptotic cells (Nardai et al. 2000). HSP90 also aids the vascular endothelial growth factor (VEGF)-induced expression of antiapoptotic Bcl-2 (Dias et al. 2002). HSP90 plays a regulatory role in endothelial and neuronal nitric oxide synthase-mediated NO production; hence, inhibition of HSP90 helps induce apoptosis by diminishing NO production and increasing NOS-dependent superoxide production in certain cellular systems (Sreedhar and Csermely 2004). HSP90 inhibition leads to the dissociation of various HSP90 client proteins from the chaperone complex and to their consecutive degradation by proteasome (An et al. 2000; Blagosklonny 2002; Schulte et al. 1997; Solit et al. 2002). HSP90 inhibition also leads to a defect in a number of proliferative signals including the Akt-dependent survival pathway (Basso et al. 2002; Munster et al. 2001, 2002). The underlying antiapoptotic mechanism proposed for the promotion of cell survival by HSPs includes the inhibition of ROS, increases in glutathione levels, and the regulation of the intracellular events and activities of a wide variety of signaling proteins.

11 Seasonal Influences on HSP Expression

Fish are ectothermic vertebrates that inhabit an aquatic environment with high temperature conductivity, and thus temperature has had an important influence on their biogeographic distribution during their evolution. In addition, periodic temperature fluctuations have an important impact on individual fish. Lejeusne et al. (2006) demonstrated that HSP50 and HSP60 expressions vary seasonally with natural temperature fluctuations in a "thermotolerant" Mediterranean marine species.

Seasonal variation may affect aquatic organisms by changing physiochemical and biological variables. Seasonal variation in the endogenous levels of HSP70 (Padmini and Usha Rani 2008) and HSP90 (Padmini and Usha Rani 2009a; Padmini et al. 2008b) has been reported in *M. cephalus* hepatocytes; such ecologically relevant seasonal variation in the expression of HSPs confers tolerance to cytotoxic effects of environmental contaminants and provides cells with stress tolerance to subsequent insults (Li and Nussenzweig 1996). There are several reports of seasonal variation in metal distribution and the influence of strong hydrodynamic and physiochemical aquatic conditions on HSP variation (Padmini and Kavitha 2003; Padmini and Vijaya Geetha 2007a; Padmini et al. 2009a). Fader et al. (1994) also reported that the sensitivity of HSP expression can vary by season.

Fluctuating concentrations of cellular stress proteins may be especially significant in the environmental adaptation of eurythermal ectotherms (Chapple et al. 1998), as endogenous levels of HSP70 in *Mytilus edulis* vary seasonally and are positively correlated with seasonal changes, both in environmental temperatures and thermal tolerance. Seasonal changes in HSPs have also been reported in species such as *Mytilus trossulus*, where higher levels of HSP70 were recorded in the gills of summer-collected animals than in the gills of winter-collected animals (Hofmann and Somero 1995). The HSP70 isoform showed the greatest seasonal changes in concentration, and its variation also correlated more closely with changes in temperature and thermotolerance than did that of HSP72 and HSP78. The HSP70 was not detectable in the winter but was strongly induced as temperatures increased in the summer. In addition, HSP70 is strongly involved in the development of seasonal thermal tolerance in *M. edulis*. Seasonal variation in the tissue levels of constitutive HSP70 has been demonstrated in four species of stream fish (*Pimephales promelas, Salmo trutta, Ictalurus natalis,* and *Ambloplites rupestris*), with the highest levels recorded in the spring, followed by the summer, fall, and winter. Dietz and Somero (1993) demonstrated that summer-acclimated gobies (genus *Gillichthys*) had significantly higher levels of brain HSP90 than winter-acclimated fish, and that the threshold induction temperature for HSP90 was 4°C higher in the former group. Dietz and Somero (1992) also suggested that higher levels of HSPs in eurythermal ectotherms during summer could provide cells with adequate ability to process partially unfolded proteins; in the monsoon season, heavy rains dilute pollutants and thus protein denaturation may occur less frequently and low concentrations of HSPs may be adequate. Collectively, these data suggest that HSP expression, which is subject to acclimatization, correlates well with seasonal changes in thermal tolerance in aquatic environments.

12 Conclusions

1. Fish that inhabit polluted environments exhibit organ-specific alterations in response to stress, which results in accumulation of denatured or partially unfolded cellular proteins.

2. To counteract stress-specific damage, fish trigger the overexpression of HSPs as one of their adaptive responses. The upregulation of HSPs plays a key role in protecting fish from environmental stress and stress-induced reductions in viability, thereby enhancing their survival.
3. This response of HSPs to stressors will potentially serve as a useful future biomarker of environmental quality for aquatic organisms. Analysis of HSF1, the precursor for HSP synthesis, aids in accessing the influence of external factors on HSP expression. Further studies on HSP expression, using other fish models under natural field conditions, will also support the usage of HSF1 as a biomarker.
4. The impact of seasonal influences on constitutive and inducible factors, such as HSP expression, is a particularly important line of future research, because seasonal variation in environmental parameters has a major effect on the cellular homeostasis of organisms.

13 Summary

Fish are subjected to a wide variety of environmental stressors. Stressors affect fish at all life stages and the stress-specific responses that occur at the biochemical and physiological levels affect the overall health and longevity of such animals. In this review, the organ-specific alterations in fish that inhabit polluted environments are addressed in detail. Fish, like other vertebrates, have evolved strategies to counteract stress-mediated effects. Among the key strategies that fish have developed is the induction of HSPs. The primary functions of HSPs are to promote the proper folding or refolding of proteins, to prevent potentially damaging interactions with proteins, and aiding in the disassembly of formations of protein aggregates.

Stress, a state of unbalanced tissue oxidation, causes a general disturbance in the cellular antioxidant and redox balance and evokes HSP70 overexpression. Distinct families of HSPs have diverse physiological functions, and their induction, which is regulated at the transcriptional level, is mediated by the activation of heat shock factors. Interestingly, HSPs also interact with a wide variety of signaling molecules that modulate stress-mediated apoptotic effects. Hence, HSP induction is of major importance for maintenance of cell homeostasis. HSP-mediated adaptation processes are regarded as a fundamental protective mechanism that decreases cellular sensitivity to damaging events. Thus, the adaptive expression of HSPs is a protective response that helps combat stress-induced conformational damage to proteins.

Additional research is needed to gain further information on the functional significance and role of individual HSPs and to enhance the understanding of the molecular mechanisms by which they act. In addition, field studies are needed to allow comprehensive evaluation of the potential use of HSPs as biomarkers for environmental monitoring. Furthermore, the expression of HSPs in fish fluctuates in response to seasonal variation. Because HSPs serves as a tool for assessing the stressed state of individuals and/or populations, the impact of seasonal influences on constitutive and inducible factors of these proteins should also be elucidated. Such research will lead to a fundamental improvement in the understanding of the functional role of HSPs in response to natural environmental changes and may allow correlation of the action of HSPs at the molecular level with the whole organismal stress response, which, so far, remains unexplained.

Acknowledgment The Department of Science and Technology (DST) project (SP/SO/AS-10/2003) funded by Ministry of Science and Technology, Government of India is acknowledged.

References

Abele D, Puntarulo S (2004) Formation of reactive species and induction of antioxidant defense systems in polar and temperate marine invertebrates and fish. Comp Biochem Physiol B 138:405–415

Ackerman PA, Forsyth RB, Mazur CF, Iwama GK (2000) Stress hormones and the cellular stress response in salmonids. Fish Physiol Biochem 23:327–336

Ahmad I, Hamid T, Fatima M, Chand HS, Jain SK, Athar M, Raisuddin S (2000) Induction of hepatic antioxidants in freshwater catfish (*Channa punctatus* Bloch.) is a biomarker of paper mill effluent exposure. Biochim Biophys Acta 1523:37–48

Airaksinen S, Rabergh CM, Sistonen L, Nikinmaa M (1998) Effects of heat shock and hypoxia on protein synthesis in rainbow trout (*Oncorhynchus mykiss*) cells. J Exp Biol 201: 2543–2551

Airaksinen S, Rabergh CMI, Lahti A, Kaatrasalo A, Sistonen L, Nikinmaa M (2003) Stressor-dependent regulation of the heat shock response in Zebrafish, *Danio rerio*. Comp Biochem Physiol A 134:839–846

An WG, Schulte TW, Neckers LM (2000) The heat shock protein 90 antagonist geldanamycin alters chaperone association with p210$^{bcr-abl}$ and v-src proteins before their degradation by the proteasome. Cell Growth Differ 11:355–360

Arai A, Naruse K, Mitani H, Shima A (1995) Cloning and characterization of cDNAs for 70-kDa heat-shock proteins (Hsp70) from two fish species of the genus *Oryzias*. Jpn J Genet 70: 423–433

Arrigo AP (2001) HSP27: novel regulator of intracellular redox state. IUBMB Life 52:303–307

Barja G, Herrero A (2000) Oxidative damage to mitochondrial DNA is inversely related to maximum life span in the heart and brain of mammals. FASEB J 14:312–318

Barton BA (2002) Stress in fishes: a diversity of responses with particular reference to changes in circulating corticosteroids. Integr Comp Biol 42:517–525

Barton BA, Morgan JD, Vijayan MM (2002) Physiological and condition-related indicators of environmental stress in fish. In: Adams SM (ed) Biological indicators of aquatic ecosystem stress. American Fisheries Society, Bethesda, MD, pp 111–148

Basso AD, Solit DB, Chiosis G, Giri B, Tsichlis P, Rosen N (2002) Akt forms an intracellular complex with heat shock protein 90 (HSP90) and Cdc37 and is destabilized by inhibitors of HSP90 function. J Biol Chem 277:39858–39866

Basu N, Nakano T, Grau EG, Iwama GK (2001) The effects of cortisol on heat shock protein 70 levels in two fish species. Gen Comp Endocrinol 124:97–105

Basu N, Todgham AE, Ackerman PA, Bibeau MR, Nakano K, Schulte PM, Iwama GK (2002) Heat shock protein genes and their functional significance in fish. Gene 295:173–183

Beere HM (2004) 'The stress of dying': the role of heat shock proteins in the regulation of apoptosis. J Cell Sci 117:2641–2651

Beere HM, Wolf BB, Cain K, Mosser DD, Mahboubi A, Kuwana T, Tailor P, Morimoto RI, Cohen GM, Green DR (2000) Heat shock protein 70 inhibits apoptosis by preventing recruitment of procaspase-9 to the Apaf-1 apoptosome. Nat Cell Biol 2:469–475

Benjamin IJ, McMillan DR (1998) Stress (heat shock) proteins: molecular chaperones in cardiovascular biology and disease. Circ Res 83:117–132

Blagosklonny MV (2002) HSP90 associated oncoproteins: multiple targets of geldanamycin and its analogs. Leukemia 16:455–462

Brown MA, Upender RP, Hightower LE, Renfro JL (1992) Thermoprotection of a functional epithelium: heat stress effects on transepithelial transport by flounder renal tubule in primary monolayer culture. Proc Natl Acad Sci USA 89:3246–3250

Bruey JM, Ducasse C, Bonniaud P, Ravagnan L, Susin SA, Diaz-Latoud C, Gurbuxani S, Arrigo AP, Kroemer G, Solary E, Garrido C (2000) HSP27 negatively regulates cell death by interacting with cytochrome *c*. Nat Cell Biol 2:645–652

Buchner J (1999) HSP90 and co – a holding for folding. Trends Biochem Sci 24:136–141

Buckley BA, Place SP, Hofmann GE (2004) Regulation of heat shock genes in isolated hepatocytes from an Antarctic fish, *Trematomus bernacchii*. J Exp Biol 207:3649–3656

Buet A, Barilloet S, Camilleri V (2002) Changes in oxidative stress parameters in fish as response to direct uranium exposure. Radioprot Suppl 40:S151–S155

Buettner GR, Schafer FQ (2000) Free radicals, oxidants and antioxidants. Teratol 62:234

Bukau B, Horwich AL (1998) The HSP70 and HSP60 chaperone machines. Cell 92:351–366

Calabrese V, Scapagnini G, Ravagna A, Fariello RG, Giuffrida Stella AM, Abraham NG (2002) Regional distribution of heme oxygenase, HSP70, and glutathione in brain: relevance for endogenous oxidant/antioxidant balance and stress tolerance. J Neurosci Res 68: 65–75

Cappo M, Alongi DM, Williams D, Duke N (1998) A review and synthesis of Australian Fisheries Habitat Research: major threats, issues and gaps in knowledge of coastal and marine fisheries habitats. Fish Res Dev Corp. pp. 53

Cara JB, Aluru N, Moyano FJ, Vijayan MM (2005) Food-deprivation induces HSP70 and HSP90 protein expression in larval gilthead sea bream and rainbow trout. Comp Biochem Physiol B 142:426–431

Carpenter CM, Hofmann GE (2000) Expression of 70 kDa heat shock proteins in Antarctic and New Zealand notothenioid fish. Comp Biochem Physiol A Physiol 125:229–238

Carratu L, Gracey AY, Buono S, Maresca B (1998) Do Antarctic fish respond to heat shock? In: di Prisco G, Pisano E, Clarke A (eds) Fishes of the Antarctic: a biological overview. Springer, Milan, pp 111–118

Chapple JP, Smerdon GR, Berry RJ, Hawkins AJS (1998) Seasonal changes in stress-70 protein levels reflect thermal tolerance in the marine bivalve *Mytilus edulis* L. J Exp Mar Biol Ecol 229:53–68

Chauhan D, Li G, Hideshima T, Podar K, Mitsiades C, Mitsiades N, Catley L, Tai YT, Hayashi T, Shringharpure R, Burger R, Munshi N, Ohtake Y, Saxena S, Anderson KC (2003) HSP27 inhibits release of mitochondrial protein Smac in multiple myeloma cells and confers dexamethasone resistance. Blood 102:3379–3386

Chen Z, Cheng C.-HC, Zhang J, Cao L, Chen L, Zhou L, Jin Y, Ye H, Deng C, Dai Z, Xu Q, Hu P, Sun S, Shen Y, Chen L (2008) Transcriptomic and genomic evolution under constant cold in Antarctic notothenioid fish. Prot Natl Acad Sci USA 105:12944–12949

Cheng P, Liu X, Zhang G, Deng Y (2006) Heat shock protein 70 gene expression in four hatchery Pacific Abalone *Haliotis discus hannai* Ino populations using for marker-assisted selection. Aquaculture Res 37:1290–1296

Cissel DS, Beaven MA (2000) Disruption of Raf-1/heat shock protein 90 complex and Raf signaling by dexamethasone in mast cells. J Biol Chem 275:7066–7070

Clarkson K, Kieffer JD, Currie S (2005) Exhaustive exercise and the cellular stress response in rainbow trout, *Oncorhynchus mykiss*. Comp Biochem Physiol A 140:225–232

Cossu C, Doyotte A, Jacquin MC, Babut M, Exinger A, Vasseur P (1997) Glutathione reductase, selenium-dependent glutathione peroxidase, glutathione levels and lipid peroxidation in freshwater bivalves, *Unio tumidus*, as biomarkers of aquatic contamination in field studies. Ecotoxicol Environ Saf 38:122–131

Cossu C, Doyotte A, Babut M, Exinger A, Vasseur P (2000) Antioxidant biomarkers in freshwater bivalves, *Unio tumidus*, in response to different contamination profiles of aquatic sediments. Ecotoxicol Environ Saf 45:106–121

Csermely P (2004) Strong links are important, but weak links stabilize them. Trends Biochem Sci 29:331–334

Csermely P, Schnaider T, Soti C, Prohaszka Z, Nadai G (1998) The 90-kDa molecular chaperone family: structure, function and clinical applications. A comprehensive review. Pharmacol Ther 79:129–168

Currie S, Tufts BL (1997) Synthesis of stress protein 70 (HSP70) in rainbow trout (*Oncorhynchus mykiss*) of red blood cells. J Exp Biol 200:607–614

Currie S, Tufts BL, Moyes CD (1999) Influence of bioenergetic stress on heat shock protein gene expression in nucleated red blood cells of fish. Am J Physiol Regul Integr Comp Physiol 276:R990–R996

Currie S, Moyes CD, Tufts BL (2000) The effects of heat shock and acclimation temperature on hsp70 and hsp30 mRNA expression in rainbow trout: in vivo and in vitro comparisons. J Fish Biol 56:398–408

Dautremepuits C, Marcogliese DJ, Gendron AD, Fournier M (2009) Gill and head kidney antioxidant processes and innate immune system responses of yellow perch (*Perca flavescens*) exposed to different contaminants in the St. Lawrence River, Canada. Sci Total Environ 407: 1055–1064

Delaney MA, Klesius PH (2004) Hypoxic conditions induce Hsp70 production in blood, brain and head kidney of juvenile Nile tilapia *Oreochromis niloticus* (L.). Aquaculture 236:633–644

Demand J, Alberte S, Patterson C, Höhfeld J (2001) Cooperation of a ubiquitin domain protein and an E3 ubiquitin ligase during chaperone/proteasome coupling. Curr Biol 11:1569–1577

Dhaliwal GS, Kukal SS (2005) Essentials of environmental science. Kalyani Publishers, Ludhiana, India, pp 224–234

Dias S, Shmelkov SV, Lam G, Rafii S (2002) VEGF (165) promotes survival of leukemic cells by HSP90-mediated induction of Bcl-2 expression and apoptosis inhibition. Blood 99:2532–2540

Dietz TJ, Somero GN (1992) The threshold induction temperature of the 90 kDa heat shock protein is subject to acclimatization in eurythermal goby fishes (Genus *Gillichthys*). Proc Natl Acad Sci USA 89:3389–3393

Dietz TJ, Somero GN (1993) Species- and tissue-specific synthesis patterns for heat shock proteins HSP70 and HSP90 in several marine teleost fishes. Physiol Zool 66:863–880

Doyotte AC, Jacquin MC, Babut M, Vasseur P (1997) Antioxidant enzymes, glutathione and lipid peroxidation of experimental or field exposure in the gills and the digestive gland of the freshwater bivalve *Unio tumidus*. Aquat Toxicol 39:93–110

Dragovic Z, Broadley SA, Shomura Y (2006) Molecular chaperones of the HSP110 family act as nucleotide exchange factors of HSP70s. EMBO J 25:2519–2528

Dubeau SF, Pan F, Tremblay GC, Bradley TM (1998) Thermal shock of salmon in vivo induces the heat shock protein (HSP70) and confers protection against osmotic shock. Aquaculture 168:311–323

Duffy LK, Scofield E, Rodgers T, Patton M, Bowyer RT (1999) Comparative baseline levels of mercury, HSP70 and HSP60 in subsistence fish from the Yukon–Kuskokwim delta region of Alaska. Comp Biochem Physiol Toxicol Pharmacol 124:181–186

Eufemia NA, Collier TK, Stein JE, Watson DE, Di Giulio RT (1997) Biochemical responses to sediment-associated contaminants in brown bullhead (*Ameiurus nebulosus*) from the Niagara River ecosystem. Ecotoxicology 6:13–34

Eustace BK, Jay DG (2004) Extracellular roles for the molecular chaperone, HSP90. Cell Cycle 3:1098–1100

Fader SC, Yu Z, Spotila JR (1994) Seasonal variation in heat shock proteins (HSP70) in stream fish under natural conditions. J Therm Biol 19:335–341

Falciani F, Diab AM, Sabine V, Williams TD, Ortega F, George SG, Chipman JK (2008) Hepatic transcriptomic profiles of European flounder (*Platichthys flesus*) from field sites and computational approaches to predict site from stress gene responses following exposure to model toxicants. Aquat Toxicol 90:92–101

Farkas T, Kutskova YA, Zimarino V (1998) Intramolecular repression of mouse heat shock factor1. Mol Cell Biol 18:906–918

Feder ME, Hofmann GE (1999) Heat-shock proteins, molecular chaperones, and the stress response: evolutionary and ecological physiology. Annu Rev Physiol 61:243–282

Feder ME, Krebs RA (1998) Natural and genetic engineering of thermotolerance in *Drosophila melanogaster*. Am Zool 38:503–517

Fedoroff N (2006) Redox regulatory mechanisms in cellular stress responses. Ann Bot 98:289–300

Fenet H, Casellas C, Bontoux J (1996) Hepatic enzymatic activities of the European eel *Anguilla anguilla* as a tool for biomonitoring fresh-water streams: laboratory and field caging studies. Water Sci Technol 33:321–329

Fernandes M, O'Brien T, Lis JT (1994) Structure and regulation of heat shock gene promoters. In: Morimoto RI, Tissieres A, Georgopoulos C (eds) The biology of heat shock proteins and molecular chaperones. Cold Spring Harbor Laboratory Press, Cold Spring Harbor, NY, pp 375–393

Fink AL, Goto Y (1998) Molecular chaperones in the life cycle of proteins: structure, function and mode of action. Marcel Dekker, New York, NY

Freeman ML, Spitz DR, Meredith MJ (1990) Does heat shock enhance oxidative stress? Studies with ferrous and ferric iron. Radiat Res 124:288–293

Gabai VL, Sherman MY (2002) Interplay between molecular chaperones and signaling pathways in survival of heat shock. J Appl Physiol 92:1743–1748

Gamperl AK, Vijayan MM, Boutilier RG (1994) Experimental control of stress hormone levels in fishes: techniques and applications. Rev Fish Biol Fish 4:215–255

Gotoh T, Terada K, Mori M (2001) hsp70-DnaJ chaperone pairs prevent nitric oxide-mediated apoptosis in RAW 264.7 macrophages. Cell Death Differ 8:357–366

Grosvik BE, Goksoyr A (1996) Biomarker protein expression in primary cultures of salmon (*Salmo salar* L) hepatocytes exposed to environmental pollutants. Biomarkers 1:45–53

Haendeler J, Hoffmann J, Rahman S, Zeiher AM, Dimmeler S (2003) Regulation of telomerase activity and anti-apoptotic function by protein–protein interaction and phosphorylation. FEBS Lett 536:180–186

Hartl FU (1996) Molecular chaperones in cellular protein folding. Nature 381:571–579

Hartl FU, Hayer-Hartl M (2002) Molecular chaperones in the cytosol: from nascent chain to folded protein. Science 295:1852–1858

Hassanein HMA, Banhawy MA, Soliman FM, Abdel-Rehim SA, Muller WEG, Schroder HC (1999) Induction of HSP70 by the herbicide oxyfluorfen (goal) in the Egyptian Nile fish, *Oreochromis niloticus*. Arch Environ Contam Toxicol 37:78–84

Hernandez C, Martin M, Bodega G, Suarez I, Perez J, Fernandez B (1999) Response of carp central nervous system to hyperammonemic conditions: an immunocytochemical study of glutamine synthetase (GS), glial fibrillary acidic protein (GFAP) and 70 kDa heat-shock protein (HSP70). Aquat Toxicol 45:195–207

Hofmann G, Buckley BA, Airaksinen S, Keen JE, Somero GN (2000) Heat shock protein expression is absent in the Antarctic fish *Trematomus bernachhii* (family Nototheniidae). J Exp Biol 203:2331–2339

Hofmann GE, Somero GN (1995) Evidence for protein damage at environmental temperatures: seasonal changes in levels of ubiquitin conjugates and HSP70 in the intertidal mussel *Mytilus trossulus*. J Exp Biol 198:1509–1518

Iwama GK, Afonso LOB, Todgham A, Ackerman P, Nakano K (2004) Are hsps suitable for indicating stressed states in fish? J Exp Biol 207:15–19

Iwama GK, Thomas PT, Forsyth RB, Vijayan MM (1998) Heat shock protein expression in fish. Rev Fish Biol Fish 8:35–56

Iwama GK, Vijayan MM, Forsyth RB, Ackerman PA (1999) Heat shock proteins and physiological stress in fish. Am Zool 39:901–909

Ju Z, Karsi A, Kocabas A, Patterson A, Li P, Cao D, Dunham R, Liu Z (2000) Transcriptome analysis of channel catfish (*Ictalurus punctatus*): genes and expression profile from the brain. Gene 261:373–382

Keaney M, Matthijssens F, Sharpe M, Vanfletern J, Gems D (2004) Superoxide dismutase mimetics elevate superoxide dismutase activity in vivo but do not retard aging in the nematode *Caenorhabditis elegans*. Free Radic Biol Med 37:239–250

Knowlton AA, Gupta S (2003) HSP60, Bax, and cardiac apoptosis. Cardiovasc Toxicol 3:263–268

Kong HJ, Kang HS, Kim HD (1996) Expression of the heat shock proteins in HeLa and CHSE-214 cells exposed to heat shock. Korean J Zool 39:123–131

Kothary RK, Burgess EA, Candido EPM (1984) The heat-shock phenomenon in cultured cells of rainbow trout: hsp70 mRNA synthesis and turnover. Biochim Biophys Acta 783:137–143

Krone PH, Lele Z, Sass JB (1997) Heat shock genes and the heat shock response in zebrafish embryos. Biochem Cell Biol 75:487–497

Krone PH, Sass JB (1994) *Hsp90α* and *hsp90β* genes are present in the zebrafish and are differentially regulated in developing embryos. Biochem Biophys Res Commun 204:746–752

Lejeusne C, Perez T, Sarrazin V, Chevaldonne P (2006) Baseline expression of HSPs of a 'thermotolerant' Mediterranean marine species largely influenced by natural temperature fluctuations. Can J Fish Aquat Sci 63:2028–2037

Lele Z, Engel S, Krone PH (1997) Hsp47 and hsp70 gene expression is differentially regulated in a stress- and tissue-specific manner in zebrafish embryos. Dev Genet 21:123–133

Lenaire P, Livingstone DR (1993) Pro-oxidant/antioxidant processes and organic xenobiotic interactions in marine organisms, in particular the flounder *Platichthys flesus* and the mussel *Mytilus edulis*. Trends Comp Biochem Physiol 1:1119–1150

Li GC, Nussenzweig A (1996) Thermotolerance and heat shock proteins: possible involvement of Ku autoantigen in regulating HSP70 expression. In: Feige U, Morimoto RI, Yahara I, Polla B (eds) Stress-inducible cellular responses. Birkhäuser Verlag, Basel, pp 425–449

Li C-Y, Lee J-S, Ko Y-G, Kim J-I, Seo J-S (2000) Heat shock protein 70 inhibits apoptosis downstream of cytochrome *c* release and upstream of caspase-3 activation. J Biol Chem 275:25665–25671

Lim EH, Brenner S (1999) Short-range linkage relationships, genomic organization and sequence comparisons of a cluster of five hsp70 genes in *Fugu rubripes*. Cell Mol Life Sci 55:668–678

Lindquist S (1986) The heat-shock response. Annu Rev Biochem 55:1151–1191

Lund SG, Caissie D, Cunjak RA, Vijayan MM, Tufts BL (2002) The effects of environmental heat stress on heat-shock mRNA and protein expression in Miramichi Atlantic salmon (*Salmo salar*) parr. Can J Fish Aquat Sci 59:1553–1562

Martin M, Hernandez C, Bodega G, Suarez I, Boyano MC, Fernandez B (1998) Heat-shock proteins expression in fish central nervous system and its possible relation with water acidosis resistance. Neurosci Res 31:97–106

Martin CC, Tang P, Barnardo G, Krone PH (2001) Expression of the chaperonin 10 gene during zebrafish development. Cell Stress Chaperones 6:38–43

Mearow KM, Dodge ME, Rahimtula M, Yegappan C (2002) Stress-mediated signaling in PC12 cells – the role of the small heat shock protein, HSP27, and Akt in protecting cells from heat stress and nerve growth factor withdrawal. J Neurochem 83:452–462

Molina A, Biemar F, Muller F, Iyengar A, Prunet P, Maclean N, Martial JA, Muller M (2000) Cloning and expression analysis of an inducible HSP70 gene from tilapia fish. FEBS Lett 474:5–10

Morange M (1997) Developmental control of heat shock and chaperone gene expression. Cell Mol Life Sci 53:78–79

Morimoto RI, Santoro MG (1998) Stress-inducible responses and heat shock proteins: new pharmacologic targets for cytoprotection. Nature Biotechnol 16:833–838

Morimoto RI, Sarge KD, Abravaya K (1992) Transcriptional regulation of heat shock genes. A paradigm for inducible genomic responses. J Biol Chem 267:21987–21990

Mosser DD, Bols NC (1988) Relationship between heat-shock protein synthesis and thermotolerance in rainbow trout fibroblasts. J Comp Physiol B 158:457–467

Munster PN, Basso A, Solit D, Norton L, Rosen N (2001) Modulation of HSP90 function by ansamycins sensitizes breast cancer cells to chemotherapy-induced apoptosis in an RB- and schedule-dependent manner. Clin Cancer Res 7:2228–2236

Munster PN, Marchion DC, Basso AD, Rosen N (2002) Degradation of HER2 by ansamycins induces growth arrest and apoptosis in cells with HER2 overexpression via a HER3, phosphatidylinositol 3′-kinase-AKT-dependent pathway. Cancer Res 62:3132–3137

Nakano K, Iwama GK (2002) The 70-kDa heat shock protein response in two intertidal sculpins, *Oligocottus maculosus* and *O. snyderi*: relationship of hsp70 and thermal tolerance. Comp Biochem Physiol A 133:79–94

Nardai G, Sass B, Eber J, Orosz G, Csermely P (2000) Reactive cysteines of the 90-kDa heat shock protein, Hsp90. Arch Biochem Biophys 384:59–67

Niu CJ, Rummer JL, Brauner CJ, Schulte PM (2008) Heat shock protein (Hsp70) induced by a mild heat shock slightly moderates plasma osmolarity increases upon salinity transfer in rainbow trout (*Oncorhynchus mykiss*). Comp Biochem Physiol C 148:437–444

Nollen EAA, Morimoto RI (2002) Chaperoning signaling pathways: molecular chaperones as stress-sensing 'heat shock' proteins. J Cell Sci 115:2809–2816

Norris CE, Brown MA, Hickey E, Weber LA, Hightower LE (1997) Low-molecular weight heat shock proteins in a desert fish (*Poeciliopsis lucida*): Homologs of human *hsp27* and *Xenopus hsp30*. Mol Cell Evol 14:1050–1061

Orr WC, Sohal RS (1994) Extension of life-span by overexpression of superoxide dismutase and catalase in *Drosophila melanogaster*. Science 263:1128–1130

Padmini E, Kavitha M (2003) Seasonal pollution assessment through comparative hydrobiological studies in Ennore and Kovalam estuaries. Indian Hydrobiol 6:139–144

Padmini E, Kavitha M (2005a) Contaminant induced stress impact on the histology and biochemical alterations in the brain of estuarine grey mullets. Pollut Res 24:177–181

Padmini E, Kavitha M (2005b) Evaluation of genotoxic effects due to contaminant mediated oxidative damage in the brain of *Mugil cephalus* (Linnaeus). Pollut Res 24:33–36

Padmini E, Kavitha M (2007) Comparative assessment of contaminant induced oxidative stress in brain of *Mugil cephalus*. Environ Pollut Control J 10:75–79

Padmini E, Sudha D (2004) Environmental impact on gill mitochondrial function in *Mugil cephalus*. Aquaculture 5:89–92

Padmini E, Usha Rani M (2008) Impact of seasonal variation on HSP70 expression quantitated in stressed fish hepatocytes. Comp Biochem Physiol B 151:278–285

Padmini E, Usha Rani M (2009a) Seasonal influence on heat shock protein 90α and heat shock factor 1 expression during oxidative stress in fish hepatocytes from polluted estuary. J Exp Mar Biol Ecol 372:1–8

Padmini E, Usha Rani M (2009b) Evaluation of oxidative stress biomarkers in hepatocytes of grey mullet inhabiting natural and polluted estuaries. Sci Total Environ 407:4533–4541

Padmini E, Usha Rani M (2010) Thioredoxin and HSP90α modulate ASK1-JNK1/2 signaling in stressed hepatocytes of *Mugil cephalus*. Comp Biochem Physiol C 151:187–193

Padmini E, Vijaya Geetha B (2007a) Seasonal influences on water quality parameters and pollution status of the Ennore estuary, Tamilnadu, India. J Environ Hydrol 15:1–15

Padmini E, Vijaya Geetha B (2007b) A comparative seasonal pollution assessment study on Ennore estuary with respect to metal accumulation in the grey mullet, *Mugil cephalus*. Oceanol Hydrobiol Stud 35:91–103

Padmini E, Vijaya Geetha B (2007c) Oxidative stress biomarkers in the liver mitochondria of grey mullets shows seasonal variation in the Ennore estuary. J Ecophysiol Occup Health 7:107–116

Padmini E, Vijaya Geetha B (2008) Biodegradative efficiency of recombinant *Escherichia coli* on heavy metal contamination and organic pollutants from Ennore estuary. Asian J Microbiol Biotechnol Environ Sci 14:185–190

Padmini E, Vijaya Geetha B (2009a) Impact of season on liver mitochondrial oxidative stress and the expression of HSP70 in grey mullets from contaminated estuary. Ecotoxicology 18:304–311

Padmini E, Vijaya Geetha B (2009b) Modulation of ASK1 expression during overexpression of Trx and HSP70 in stressed fish liver mitochondria. Cell Stress Chaperones 14:459–467

Padmini E, Sridevi S, Vijaya Geetha B (2006) Environmental stress in Ennore estuary and enhanced erythrocyte micronuclei formation in mullets. Environ Pollut Control J 9:51–56

Padmini E, Thendral Hepsibha B, Shanthalin Shellomith AS (2004) Lipid alteration as stress markers in grey mullets (*Mugil cephalus* Linnaeus) caused by industrial effluents in Ennore estuary. Aquaculture 5:115–118

Padmini E, Vijaya Geetha B, Usha Rani M (2008a) Liver oxidative stress of the grey mullet *Mugil cephalus* presents seasonal variations in Ennore estuary. Braz J Med Biol Res 41:951–955

Padmini E, Usha Rani M, Vijaya Geetha B (2008b) Differential HSP90α expression in fish hepatocytes from polluted estuary during summer. Fish Sci 74:1118–1126

Padmini E, Usha Rani M, Vijaya Geetha B (2009a) Studies on antioxidant status in *Mugil cephalus* in response to heavy metal pollution at Ennore estuary. Environ Monit Assess 155:215–225

Padmini E, Vijaya Geetha B, Usha Rani M (2009b) Pollution induced nitrative stress and heat shock protein 70 overexpression in fish liver mitochondria. Sci Total Environ 407:1307–1317

Palmisano AN, Winton JR, Dickhoff, WW (2000) Tissue specific induction of *hsp90* mRNA and plasma cortisol response in Chinook salmon following heat shock, seawater challenge, and handling challenge. Mar Biotechnol 2:329–338

Pan F, Zarate JM, Tremblay GC, Bradley TM (2000) Cloning and characterization of salmon *hsp90* cDNA: upregulation by thermal and hyperosmotic stress. J Exp Zool 287:199–212

Pandey P, Saleh A, Nakazawa A, Kumar S, Srinivasula SM, Kumar V, Weichselbaum R, Nalin C, Alnemri ES, Kufe D, Kharbanda S (2000) Negative regulation of cytochrome *c*-mediated oligomerisation of Apaf-1 and activation of procaspase-9 by heat shock protein 90. EMBO J 19:4310–4322

Pandey S, Ahmad I, Parvez S, Bin-Hafeez B, Haque R, Raisuddin S (2001) Effect of endosulfan on antioxidants of freshwater fish *Channa punctatus* Bloch: 1. Protection against lipid peroxidation in liver by copper preexposure. Arch Environ Contam Toxicol 41:345–352

Pandey S, Parvez S, Sayeed I, Haque R, Bin-Hafeez B, Raisuddin S (2003) Biomarkers of oxidative stress: a comparative study of river Yamuna fish *Wallago attu* (Bl. & Schn.). Sci Total Environ 309:105–115

Papp E, Nardai G, Soti C, Csermely P (2003) Molecular chaperones, stress proteins and redox homeostasis. Biofactors 17:249–257

Parihar MS, Javeri T, Hemnani T, Dubey AK, Prakash P (1997) Responses of superoxide dismutase, glutathione peroxidase and reduced glutathione antioxidant defenses in gills of the freshwater catfish (*Heteropneustes fossilis*) to short-term elevated temperature. J Therm Biol 22:151–156

Parker JD, Parker KM, Sohal BH, Sohal RS, Keller L (2004) Decreased expression of Cu–Zn superoxide dismutase 1 in ants with extreme lifespan. Proc Natl Acad Sci USA 101:3486–3489

Parsell DA, Lindquist S (1993) The function of heat-shock proteins in stress tolerance: degradation and reactivation of damaged proteins. Annu Rev Genet 27:437–496

Parsell DA, Lindquist S (1994) Heat shock proteins and stress tolerance. In: Morimoto RI, Tissieres A, Georgopoulos C (eds) Biology of heat shock proteins and molecular chaperones. Cold Spring Harbor Laboratory Press, Cold Spring Harbor, NY, pp 457–494

Pearson DS, Kulyk WM, Kelly GM, Krone PH (1996) Cloning and characterization of a cDNA encoding the collagen-binding stress proteins *Hsp47* in zebrafish. DNA Cell Biol 15:263–271

Piatelli MJ, Doughty C, Chiles TC (2002) Requirement for a hsp90 chaperone-dependent MEK1/2-ERK pathway for B cell antigen receptor-induced cyclin D2 expression in mature B lymphocytes. J Biol Chem 277:12144–12150

Picard D (2002) Heat-shock protein 90, a chaperone for folding and regulation. Cell Mol Life Sci 59:1640–1648

Pirkkala L, Nykanen P, Sistonen L (2001) Roles of the heat shock transcription factors in regulation of the heat shock response and beyond. FASEB J 15:1118–1131

Poltronieri C, Maccatrozzo L, Simontacchi C, Bertotto D, Funkenstein B, Patruno M, Radaelli G (2007) Quantitative RT-PCR analysis and immunohistochemical localization of HSP70 in sea bass *Dicentrarchus labrax* exposed to transport stress. Eur J Histochem 51:125–135

Rabergh CMI, Airaksinen S, Soitamo A, Bjorklund HV, Johansson T, Nikinmaa M, Sistonen J (2000) Tissue-specific expression of zebrafish (*Danio rerio*) heat shock factor 1 mRNAs in response to heat stress. J Exp Biol 203:1817–1824

Raviol H, Sadlish H, Rodriguez F, Mayer MP, Bukau B (2006) Chaperone network in the yeast cytosol: HSP110 is revealed as an HSP70 nucleotide exchange factor. EMBO J 25: 2510–2518

Rodriguez-Ariza A, Peinado J, Pueyo C, Lopez-Barea J (1993) Biochemical indicators of oxidative stress in fish from polluted littoral areas. Can J Fish Aquat Sci 50:2568–2573

Rohde M, Daugaard M, Jensen MH, Helin K, Nylandsted J, Jattela M (2005) Members of the heat-shock protein 70 family promote cancer cell growth by distinct mechanisms. Genes Dev 19:570–582

Russotti G, Brieva TA, Toner M, Yarmush ML (1996) Induction of tolerance to hyperthermia by previous heat shock using human fibroblasts in culture. Cryobiology 33:567–580

Ryan JA, Hightower LE (1994) Evaluation of heavy-metal ion toxicity in fish cells using a combined stress protein and cytotoxicity assay. Environ Toxicol Chem 13:1231–1240

Ryan JA, Hightower LE (1996) Stress proteins as molecular biomarkers for environmental toxicology. In: Feige U, Morimoto RI, Yahara I, Polla B (eds) Stress-inducible cellular responses. Birkhauser Verlag, Basel, pp 411–424

Samali A, Orrenius S (1998) Heat shock proteins: regulations of stress response and apoptosis. Cell Stress Chaperones 3:228–236

Samali A, Robertson JD, Peterson E, Manero F, van Zeijl L, Paul C, Cotgreave IA, Arrigo AP, Orrenius S (2001) HSP27 protects mitochondria of thermotolerant cells against apoptotic stimuli. Cell Stress Chaperones 6:49–58

Santacruz H, Vriz S, Angelier N (1997) Molecular characterization of a heat shock cognate cDNA of zebrafish, hsc70, and developmental expression of the corresponding transcripts. Dev Genet 21:223–233

Sass JB, Weinberg ES, Krone PH (1996) Specific localization of zebrafish hsp90α mRNA to myoD-expressing cells suggests a role for hsp90α during normal muscle development. Mech Dev 54:195–204

Sathiyaa R, Campbell T, Vijayan MM (2001) Cortisol modulates hsp90 mRNA expression in primary cultures of trout hepatocytes. Comp Biochem Physiol B Biochem Mol Biol 129:679–685

Sathiyaa R, Vijayan MM (2003) Autoregulation of glucocorticoid receptor by cortisol in rainbow trout hepatocytes. Am J Physiol Cell Physiol 284:C1508–C1515

Schett G, Steiner CW, Groger M, Winkler S, Graninger W, Smolen J, Xu Q, Steiner G (1999) Activation of Fas inhibits heat-induced activation of HSF1 and up-regulation of HSP70. FASEB J 13:833–842

Schlesinger MJ, Ashburner M, Tissiers A (1982) Heat shock proteins from bacteria to man. Cold Spring Harbor Laboratory Press, Cold Spring Harbor, NY

Schriner SE, Linford NJ, Martin GM, Treuting P, Ogburn CE, Emond M, Coskun PE, Ladiges W, Wolf N, Van Remmen H, Wallace DC, Rabinovitch PS (2005) Extension of murine lifespan by overexpression of catalase targeted to mitochondria. Science 308:1909–1911

Schulte TW, Blagosklonny MV, Romanova L, Mushinski JF, Monia BP, Johnston JF, Nguyen P, Trepel J, Neckers LM (1996) Destabilization of Raf-1 by geldanamycin leads to disruption of the Raf-1-MEK-mitogen-activated protein kinase signaling pathway. Mol Cell Biol 16: 5839–5845

Schulte TW, An WG, Neckers LM (1997) Geldanamycin-induced destabilization of Raf-1 involves the proteasome. Biochem Biophys Res Commun 239:655–659

Shan YX, Liu TJ, Su HF, Samsamshariat A, Mestril R, Wang PH (2003) HSP10 and HSP60 modulate Bcl-2 family and mitochondria apoptosis signaling induced by doxorubicin in cardiac muscle cells. J Mol Cell Cardiol 35:1135–1143

Sheader DL, Williams TD, Lyons BP, Chipman JK (2006) Oxidative stress response of European flounder (*Platichthys flesus*) to cadmium determined by a custom cDNA microarray. Mar Environ Res 62:33–44

Sherry JP (2003) The role of biomarkers in the health assessment of aquatic ecosystems. Aquat Ecosys Health Manage 6:423–440

Smith TR, Tremblay GC, Bradley TM (1999) Characterization of the heat shock protein response of Atlantic salmon (*Salmo salar*). Fish Physiol Biochem 20:279–292

Snoeckx LH, Cornelussen RN, Van Nieuwenhoven FA, Reneman RS, Van Der Vusse GJ (2001) Heat shock proteins and cardiovascular pathophysiology. Physiol Rev 81:1461–1497

Solit DB, Zheng FF, Drobnjak M, Munster PN, Higgins B, Verbel D, Heller G, Tong W, Cordon-Cardo C, Agus DB, Scher HI, Rosen N (2002) 17-Allylamino-17-demethoxygeldanamycin induces the degradation of androgen receptor and HER2/neu and inhibits the growth of prostate cancer xenografts. Clin Cancer Res 8:986–993

Sorensen JG, Kristensen TN, Loeschcke V (2003) The evolutionary and ecological role of heat shock proteins. Ecol Lett 6:1025–1037

Sorensen JG, Loeschcke V (2007) Studying stress responses in the post-genomic era: its ecological and evolutionary role. J Biosci 32:447–456

Soti C, Nagy E, Giricz Z, Vigh L, Csemerley P, Ferdinandy P (2005) Heat shock proteins as emerging therapeutic targets. Br J Pharmacol 146:769–780

Sreedhar AS, Csermely P (2004) Heat shock proteins in the regulation of apoptosis: new strategies in tumor therapy. A comprehensive review. Pharmacol Ther 101:227–257

Sreedhar AS, Pardhasaradhi BV, Khar A, Srinivas UK (2002) A cross talk between cellular signaling and cellular redox state during heat-induced apoptosis in a rat histiocytoma. Free Radic Biol Med 32:221–227

Sreedhar AS, Kalmar E, Csermely P, Shen YF (2004) HSP90 isoforms: functions, expression and clinical importance. FEBS Lett 562:11–15

Stegeman JJ, Brouver M, Di Giulio RT, Forlin L, Fowler BA, Sanders BM, Van Veld PA (1992) Molecular responses to environmental contamination: enzyme and protein systems as indicators of chemical exposure and effect. In: Huggett RJ, Kimerle RA, Mehrle PM, Bergman HL (eds) Biomarker, biochemical, physiological, and histological markers of anthropogenic stress. Lewis, Boca Raton, FL, pp 235–335

Telli Karakoc F, Hewer A, Phillips DH, Gaines AF, Yuregir G (1997) Biomarkers of marine pollution observed in species of mullet living in two eastern Mediterranean harbours. Biomarkers 2:303–309

Thomson A, Hemphill D, Jeejeebhoy KN (1998) Oxidative stress and antioxidants in intestinal disease. Dig Dis 16:152–158

Tiligada E (2006) Chemotherapy: induction of stress responses. Endocr Relat Cancer 13: S115–S124

Tomanek L, Somero G (2002) Interspecific- and acclimation-induced variation in levels of heat shock proteins 70 (HSP70) and 90 (HSP90) and heat-shock transcription factor-1 (HSF1) in congeneric marine snails (genus *Tegula*): implications for regulation of HSP gene expression. J Exp Biol 205:677–685

Van der Oost R, Goksoyr A, Celander M, Heida H, Vermeulen NPE (1996) Biomonitoring of aquatic pollution with feral eel (*Anguilla anguilla*). II. Biomarkers: pollution-induced biochemical responses. Aquat Toxicol 36:189–222

Vijayan MM, Reddy PK, Leatherland JF, Moon TW (1994) The effects of cortisol on hepatocyte metabolism in rainbow trout: a study using the steroid analogue RU486. Gen Comp Endocrinol 96:75–84

Vijayan MM, Pereira C, Forsyth RB, Kennedy CJ, Iwama GK (1997) Handling stress does not affect the expression of hepatic heat shock protein 70 and conjugation enzymes in rainbow trout treated with β-naphthoflavone. Life Sci 61:117–127

Vijayan MM, Pereira C, Kruzynski G, Iwama GK (1998) Sublethal concentrations of contaminant induce the expression of hepatic heat shock protein 70 in 2 salmonids. Aquat Toxicol 40: 101–108

Wegele H, Muller L, Buchner J (2004) HSP70 and HSP90 – a relay team for protein folding. Rev Physiol Biochem Pharamacol 151:1–44

Whitesell L, Lindquist SL (2005) HSP90 and the chaperoning of cancer. Nat Rev Cancer 5: 761–772

Wilhelm Filho D, Torres MA, Tribess TB, Pedrosa RC, Soares CH (2001) Influence of season and pollution on the antioxidant defenses of the cichlid fish acara (*Geophagus brasiliensis*). Braz J Med Biol Res 34:719–726

Williams JH, Farag AM, Stansbury MA, Young PA, Bergman HL, Petersen NS (1996) Accumulation of HSP70 in juvenile and adult rainbow trout gill exposed to metal-contaminated water and/or diet. Environ Toxicol Chem 15:1324–1328

Williams TD, Diab AM, George SG, Godfrey RE, Sabine V, Conesa A, Minchin SD, Watts PC, Chipman JK (2006) Development of the GENIPOL European flounder (*Platichthys flesus*) microarray and determination of temporal transcriptional responses to cadmium at low dose. Environ Sci Technol 40:6479–6488

Williams TD, Diab AM, George SG, Sabine V, Chipman JK (2007) Gene expression responses of European flounder (*Platichthys flesus*) to 17-beta estradiol. Toxicol Lett 168:236–248

Williams TD, Diab A, Ortega F, Sabine VS, Godfrey RE, Falciani F, Chipman JK, George SG (2008) Transcriptomic responses of European flounder (*Platichthys flesus*) to model toxicants. Aquat Toxicol 90:83–91

Xanthoudakis S, Roy S, Rasper D, Hennessey T, Cassady R, Tawa P, Ruel R, Rosen A, Nicholson DW (1999) Hsp60 accelerates the maturation of pro-caspase-3 by upstream activator proteases during apoptosis. EMBO J 18:2049–2056

Yin Z, He JY, Gong Z, Lam TJ, Sin YM (1999) Identification of differentially expressed genes in Con A-activated carp (*Cyprinus carpio* L.) leucocytes. Comp Biochem Physiol B Biochem Mol Biol 124:41–50

Young JC, Moarefi I, Hartl FU (2001) HSP90: a specialized but essential protein-folding tool. J Cell Biol 154:267–273

Zafarullah M, Wisniewski J, Shworak NW, Schieman S, Misra S, Gedamu L (1992) Molecular cloning and characterization of a constitutively expressed heat shock cognate hsc71 gene from rainbow trout. Eur J Biochem 204:893–900

Zhang R, Luo D, Miao R, Bai L, Ge Q, Sessa WC, Min W (2005) HSP90-Akt phosphorylates ASK1 and inhibits ASK1-mediated apoptosis. Oncogene 24:3954–3963

Zou J, Salminien WF, Roberts SM, Voellmy R (1998) Correlation between glutathione oxidation and trimerization of heat shock factor 1, an early step in stress induction of the HSP response. Cell Stress Chaperones 3:130–141

Phytoremediation: A Novel Approach for Utilization of Iron-ore Wastes

Monalisa Mohanty, Nabin Kumar Dhal, Parikshita Patra, Bisweswar Das, and Palli Sita Rama Reddy

Contents

1	Introduction	29
2	Iron-ore Tailings	31
3	Environmental Impact and Waste Minimization	33
4	Phytoremediation: Sustainable Remediation and Utilization of Iron-ore Tailings	34
	4.1 Phytoextraction	37
	4.2 Phytovolatilization	38
	4.3 Rhizofiltration	39
	4.4 Phytostabilization	40
	4.5 Plants Species Suitable for Phytoremediation	42
5	Hyperaccumulation by Plant Species	42
6	Summary	43
	References	44

1 Introduction

Large amounts of toxic contaminants are being released to the environment around the globe from rapid urbanization and industrialization. Among such contaminants are industrial wastes and ore tailings that result from worldwide mining activities. In mining operations, during the processing of low-grade ores, significant quantities of wastes or tailings are produced. The overburden material (also known as "waste"), generated during surface mining of minerals, causes serious environmental hazards if surrounding flora and fauna are not properly protected. It has been roughly estimated that for every ton of metal extracted from ores, roughly 2–12 t of overburden materials are being removed.

M. Mohanty (✉)
Institute of Minerals and Materials Technology (CSIR), Bhubaneswar 751013, Orissa, India
e-mail: 18.monalisa@gmail.com

During the mining and processing of sulphide ores, large quantities of overburden and wastes are generated. The waste-containing metal sulphides of Cu, Pb, Zn, Cd, etc. undergo oxidation and form sulphuric acid. Therefore, wastes resulting from the mining of sulphide ore deposits are highly acidic and are toxic to the aquatic environment. When metal sulphides react with sulphuric acid, high concentrations of toxic heavy metal ions (e.g. Cu, Zn, Pb and Cd) are released into the environment in acidic mine drainage water and may devastate the local environment. Usually, the acidic waste water generated has a pH of <3, and a soluble metal content as high as 1,800 mg/L. Other chemicals that are used in waste water concentration processes of sulphide ores, such as flotation reagents, grinding aids and flocculants, may contribute to the toxicity of tailing water when released as effluents to local water bodies.

Surface runoff (Arnaez et al. 2004; Kandel et al. 2004) from erosion, tailings carryover or other waste also poses a significant environmental risk. Explosives, such as ammonium nitrate or trinitrotoluene (TNT) used during blasting of ores, are also subject to surface runoff. Contamination of surface water may also occur from transport of mined-ore material or heavy metal ions from mining machinery maintenance and repair. In addition, significant levels of suspended particulate material (SPM) may contaminate air, which results from mining activities such as blasting, transportation, ore crushing, ore beneficiation and disposal of tailings. Significant releases of metal-containing (including mercury) dusts may result from drying of the ore concentrate. All of the aforementioned wastes are present in thousands of unvegetated and exposed tailing piles throughout the world; such waste is a definite and persistent source of contamination and exposure for nearby communities.

India has large reserves of metal-bearing ore and occupies the sixth position in the world for iron-ore reserves. Therefore, India is an important iron-ore producer and exporter. However, approximately 10–15% of the iron ore mined in India is unutilized, even now, and is discarded as tailings. The tailing wastes that are called ultrafines or slimes, mainly those ore solids having a diameter of less than 150 μm, are not regarded to be useful and hence are discarded. In India, approximately 10–12 million tons of such mined ore is lost as tailings. The safe disposal or utilization of such vast mineral wealth in the form of ultrafines or slimes has remained as a major unsolved and challenging task for the Indian iron-ore industry. Inevitably, the proportion of iron-ore wastes generated will steadily increase, because the demand for iron ores will increase. Such a view is confirmed by the number of steel plants that have been planned for future construction in the state of Orissa and other parts of India. The total production of iron ore in India is expected to exceed 400 million tons within the next decade. Therefore, dealing with the environmental consequences of such enormous quantities of tailings will be a Herculean task. It is therefore imperative that state-of-the-art iron-ore mining and processing technologies be adopted to address and implement effective utilization of tailings.

Another challenge is addressing the panoply of legacy mining waste sites that now accentuate or may contribute in the future to local environmental damage or health consequences of nearby residents. Such sites must be restored for sustainable development, or, at least, secured to prevent off-site contaminant movement.

Dealing with metal toxicity at such waste sites is a major concern. The wastes and tailings from many mines contain ~1–50 g/kg of toxic and heavy metal ions, e.g., As, Cd, Cu, Mn, Fe, Pb and Zn (Boulet and Larocque 1998; Bradshaw et al. 1978; Walder and Chavez 1995). Moreover, waste piles of tailings normally contain no organic matter or macronutrients, and usually exhibit an acidic pH, although some tailings may be alkaline (Johnson and Bradshaw 1977; Krzaklewski and Pietrzykowski 2002). Therefore, tailings-waste areas normally lack soil structure and tend to support severely stressed heterotrophic microbial communities (Mendez et al. 2007; Southam and Beveridge 1992).

There has been an increasing interest in the possibility of using vegetation to remediate contaminated mining sites, such as those described above, through plant-based technology known as phytoremediation. It is our intent in this review to address phytoremediation and associated processes as they apply to iron-ore wastes and mining sites. We will show that phytoremediation is cost-effective and feasible because plants are able to slowly absorb toxins into their tissues and thereby help clean toxins from waste sites. In addition, phytostabilization, the use of plants for in situ stabilization of tailings and metal contaminants, is a feasible alternative to more costly remediation practices (Mendez and Maier 2008). Phytostabilization promotes the conversion of tailings into useful soil material capable of sustaining normal ecological plant succession. Such use of plants to slow or prevent leaching of toxic components or erosion processes actually works better than some traditional methodologies (Dong et al. 2007; Krzaklewski and Pietrzykowski 2002; Wong et al. 1998; Ye et al. 2002). The main benefit of phytostabilization technology is that wastes need not be moved, transported or otherwise disposed of. Rather, one simply introduces the appropriate plant species and gives them time to work.

2 Iron-ore Tailings

Iron ore is being beneficiated around the world to meet the raw material requirements of the iron and steel industries. Iron ore has its own peculiar mineralogical characteristics and for optimum product extraction at any site requires tailoring of the metallurgical treatment and specific beneficiation process selected for use. The choice of beneficiation technique depends on the nature of the gangue and its association with the ore structure. The prime function of beneficiation of iron ore in India is to improve the content of extracted iron and reduce the Al–Si content of the finished iron. In India, iron-ore beneficiation proceeds mainly from washing, sizing by classification, jigging and then magnetic separation. The advantage of washing is to impart better handling properties to the ores, particularly the removal of fines, which becomes sticky in the rainy season and may pose problems during transportation to steel plants. In addition, the fines, which are preferentially accumulated with silica- and alumina-bearing minerals, are being removed as washing proceeds, thereby enhancing the quality of the iron ore. A large volume of water is required for iron-ore processing. Before tailings are transported to tailing ponds for impoundment, most water is recovered for recycling by using a dewatering process that utilizes a thickener. After beneficiation, the rejected portion of the iron ore may include coarse and fine particulates in the wash water, and these particulates may form a slurry known as wet tailings. The physical and chemical nature of such wet tailing from beneficiation plants depends on the ore type and beneficiation process used. All washing plants in India utilize ponds for disposal of tailings. Such ponds conserve resources and help control pollution. In the future, when all of the existing rich iron resource is exhausted, extraction of iron from such tailing pond waste may become economically viable.

The typical beneficiation process, as adopted by one of India's magnetite ore processing plants situated at Kudremukh, involved a three-stage crushing operation, followed by spiral classification, magnetic separation and transfer to a flotation column. Unfortunately, this plant generated approximately 29,424 t of solids (as slurry) per day while beneficiating magnetite ore. As a consequence, Indian governmental environmental laws were imposed on it and the plant ceased operation. In contrast, an Indian iron-ore mine belonging to the National Mineral Development Corporation (NMDC) at Bailadila generates tailings of 2,700 t/d, which are disposed of in 7,500 m^3 of water that has a 27–30% solids content. Other characteristics of this waste slurry is that it contains heavy amounts of total dissolved solids (TDS) equal to 250–1,500 ppm; in addition, the slurry has an ore-fine content of 95% and a clay–silica content of 5%.

Laboratory characterization of iron-ore tailings or slimes has indicated that they are largely made up of extremely fine material. More than 60% of the particulates in such slimes have diameters that are <20 μm (Das et al. 1992, 1993). Moreover, the silica and alumina content of the tailings is quite high, which requires both beneficiation and agglomeration treatment prior to their use in steel making. The distribution of particle sizes in tailing slurries is solely dependent on the beneficiation process adopted. The distribution size of particulates is important, because iron-ore particles and associated total suspended solids (TSS) constitute the main water pollutants that require downstream treatment before being discharged. The extent to which iron-ore tailings are produced at different washing plants in India from iron-ore mining activities is presented in Table 1. From the foregoing, it is evident that large quantities of iron-ore slimes are annually produced in India and the iron content of such waste streams varies between 52 and 62.8% Fe. Iron-ore tailings are also contaminated with parts per million levels of heavy metal ions such as Cu, Pb, Zn, Cr, Sn, Mo and U, as well as lower levels of macronutrients. Many of these potentially toxic elements reach and become pollutants of water.

The composition of various inorganic contaminants in a typical set of different slimes is shown in Table 2. Concentrations of toxic heavy metals such as Cu, Fe, Mn, Zn, Cr, Mo, Ni and Co have been found in mine water, as well as in iron tailings. It has also been reported that high concentrations of heavy metals, viz., Cu, Fe, Mn, Zn, Cr, Mo, Ni and Co, are also found in the soils of surrounding localities. The soil concentration of metal ions at such sites varies as follows: Fe (33.2–121.5 g/L), Mn (0.39–1.39 g/L), Cr (57–204 g/L), Co (1.3–4.6 g/L), Cu (25.8–93.0 g/L), Mo

Table 1 Fe content of iron-ore slimes from mining operations produced at different washing plants in India

Washing plants	Production (t/year)	Average Fe content (%)
Daitari	0.3	60.0
Bailadilla-14	1.2	62.8
Bailadilla-5	0.5	61.2
Barsua	0.6	52.5
Kiriburu	1.6	60.4
Donimalai	1.0	57.9
Meghahatuburu	0.6	60.0
Bolani	0.4	59.8
Noamundi	0.75	58.1
Kudremukh[a]	15.0	26.6

[a]No longer in operation
t metric tons
Source: IMMT (Institute of Minerals and Materials Technology), Bhubaneswar, India (unpublished data)

Table 2 Detailed chemical composition of different iron-ore slimes

Constituents	1[a]	2	3	4	5	6	7	8
Fe	59.8	61.2	52.5	60.3	57.9	57.8	59.3	26.8
SiO_2	2.30	6.84	7.82	2.96	6.42	4.00	4.1	51.2
AlO_3	4.52	2.81	9.88	4.96	6.28	8.30	4.8	1.82
MnO	0.08	0.8	0.1	0.12	0.08	0.03	0.03	0.08
CaO	0.09	0.11	0.11	0.14	0.12	0.08	0.09	0.11
MgO	0.06	0.05	0.07	0.07	0.05	0.04	0.06	0.06
LOI	7.0	2.34	7.40	5.10	3.90	5.20	5.2	4.05

[a]Location in India: 1 Daitari, 2 Bailadilla, 3 Barsua, 4 Kiriburu, 5 Donimalai, 6 Meghahatuburu, 7 Bolani, 8 Noamundi
[b]*LOI* loss of ignition
Source: IMMT, Bhubaneswar (unpublished data)

(1.08–4.25 g/L) and Zn (15.5–55.9 g/L; Ghosh and Sen 2001). The high levels of these toxic metal ions produce an adverse effect on growth and development of plants, animals and humans. Therefore, it is essential that eco-friendly techniques are developed to reduce potentially damaging exposures to these metals.

3 Environmental Impact and Waste Minimization

In recent decades, intensive research and development efforts have been directed towards finding cost-effective and eco-compatible solutions for minimizing and/or utilizing the waste produced in iron-ore mining operations (Bandopadhyay et al. 2002; Johnson et al. 1992). Recent trends in solid waste management that employ reengineering are strategically designed to maximize utilization of waste stream components (Bandopadhyay et al. 1999, 2002; Johnson et al. 1992; Kumar and Singh 2004; Kumar et al. 2005; U.S.EPA 2003). In addition, the recycling of solid wastes, after removal of harmful contaminants and recovery of valuable components by simple physical beneficiation techniques, is also being utilized to reduce the impact of waste streams (Das et al. 2003; Kumar and Singh 2004).

In addition to reducing the load of toxic components in waste streams, sensitive and robust eco-friendly tools that are capable of detecting the effects of toxic substances in complex aquatic ecosystems are also needed (Gustavson and Waengberg 1995). One such tool that has been employed to explore the relative propensity of waste streams to cause environmental damage is the use of mesocosms. Mesocosms utilize bacteria, phytoplankton and periphytic algae in a model system setting and have been useful for testing of sediment toxicity and contamination. If properly designed, such model systems are sensitive, reliable and require modest investment. Mesocosms are potentially useful in environmental impact assessments for determining the effects of dredging and dumping activities, and subsequent disposal of dredged spoils in the region (Alden et al. 1985; Lewis et al. 2001; Word et al. 1987).

Other tests are designed to determine the toxicity and bioavailability of metals that exist in contaminated dredge spoils, sediments and resuspended sediments in the water column. Such tests are performed in the laboratory, comprise in situ sediment bioassays, or are performed in microcosm-scale systems (Balczon and Pratt 1994; Fichet et al. 1998; Hurk et al. 1997; Togna et al. 2001). One of the most used techniques for determining the environmental risk of pollutants from mining activities is to employ green plants in removal, detoxification or stabilization of mining and processing tailings. This approach is cost-effective and eco-friendly. There are plants uniquely able to tolerate and survive high heavy metal (e.g. Zn, Cd and Ni) concentrations in soils. The details of methods that rely on such plants are described below.

4 Phytoremediation: Sustainable Remediation and Utilization of Iron-ore Tailings

The conventional technologies that are employed to remediate mine tailings generally rely on physical and chemical stabilization processes. Physical stabilization entails covering mine waste with innocuous material, generally waste rock from mining operations, gravel, topsoil from adjacent sites or a clay cap to reduce wind and water erosion. These solutions are often temporary, costly and often inadequate because capping processes are impermanent (Johnson and Bradshaw 1977). Phytoremediation is an emerging alternative approach, which offers prospects for reducing costs and potentially improving the performance of tailings environmental pollution abatement.

Phytoremediation relies on green plants as means to remove polluting substances from the substrates in which they grow and the subsequent transformation of potentially toxic pollutants into harmless ones. Most conventional technologies employed in mining-waste remediation are expensive and may actually reduce soil fertility, subsequently causing negative effects on ecosystems. In contrast, phytoremediation is cost-effective, environmentally friendly and is an aesthetically pleasing alternative that is far more suitable for use in developing countries. Phytoremediation offers an environmentally attractive means for removing toxic metals from hazardous waste sites and contaminants from soil, and achieves success by relying on selected hyperaccumulator plants, and ultimately on solar energy. Phytoremediation works well under the climatic conditions extant in India and has been confirmed through scientific experimentation to work both in ex situ and in situ projects (Blaylock and Huang 2000; Cooper et al. 1999; Ghosh and Singh 2005; Huang et al. 1997). The results of in situ phytoremediation that has been performed generally support the view that reductions of pollutants in waste material are sustainable.

Metal-contaminated soil can be remediated through the application of chemical, physical and/or biological techniques (Baker and Walker 1990). Experimentation utilizing phytoaccumulator plants to clean contaminated soil has been undertaken at the Institute of Minerals and Materials Technology (IMMT, Bhubaneswar), located in east India. Phytoremediation tests have employed several plant species, to wit: tree species, Acacia (*Acacia mangium* Willd.), Shisham (*Dalbergia sissoo* Roxb.), Ashoka (*Saraca asoca* (Roxb.) de Wilde), Sal (*Shorea robusta* Gaertn.f.); vegetable species such as tomato (*Lycopersicon esculentum* Mill.) and grass species such as lemon grass (*Cymbopogon flexuosus* (Nees ex Steud.) (Wats.)) (Figs. 1, 2, and 3; IMMT, Bhubaneswar unpublished data). All plants tested for growth on iron-ore tailings have survived. Other associated testing indicated that use of synthetic chelating agents, e.g., ethylenediaminetetraacetic acid (EDTA), organic acids or diethylene triamine penta acetate (DTPA), in the phytoremediation process, increased heavy metal uptake by plants. The degree to which different plant parts of *Brassica juncea* absorbed heavy metals during the course of this experiment is presented in Table 3. Although it is clear from this study that phytoremediation can be

Fig. 1 Luxuriant growth of Lemon grass showing different treatments (*right* to *left* – garden soil (control), I:S (1:3), I:S (1:1), I:S (3:1), IOT at time of harvest (90 days after treatment; DAT)) *I* iron-ore tailings, *S* garden soil and *IOT* iron-ore tailings

Fig. 2 Growth of tree species (90 DAT) under different soil and iron-ore tailings treatment regimes. (This research performed at IMMT – Institute of Minerals and Materials Technology, Bhubaneswar, India)

Fig. 3 Growth and fruiting in tomato plants grown in 1:1 iron-ore tailings and soil (IMMT, Bhubaneswar)

Table 3 The content (mg/kg) of metals phytoaccumulated into *B. juncea* from soil

Brassica juncea	Pb	Hg	Zn	Cr	Mn	Fe	Process
Leaf	113.97	3.65	28.35	2.41	50.93	192.88	AAS
Flower	26.19	7.35	44.35	2.21	18.61	127.29	AAS
Root	7.16	3.54	25.55	0.99	6.29	134.31	AAS
Stalk	7.37	4.02	25.22	5.77	6.43	60.09	AAS
Total	147.53	18.56	123.47	11.38	82.26	514.57	

AAS atomic absorption spectrometry
Source: http://www.saneko98.com/PHYTOREMEDIATIONNEWTECHNOLOGY2006.pdf

successful, it has yet to become a commercially available technology in India. The current status of phytoremediation in the world is still in the developmental stage and more research is needed to understand and fully implement this remediation technology. But, bench-scale studies are ongoing in the United States to understand

and assist in implementation of this alternative technology. For example, in 1996, a trial in Maine on phytoremediation for removal of lead (Pb) was implemented at selected sites by Edenspace Systems, and in 1997 at another site in Trenton, New Jersey (Henry 2000).

Phytoremediation may be carried out by methods that are either ex situ or in situ. If the method employed is ex situ, the contaminated soil or waste is removed from its native site for treatment and is later returned to the restored site. Conventional ex situ methods, when applied to remediate polluted soils, rely on excavation, detoxification and/or contaminant destruction (by physical or chemical means). Such methods are designed to stabilize, solidify, immobilize, incinerate or otherwise destroy contaminants.

In contrast, in situ remediation methods are performed at the point of the contamination and do not employ excavation of contaminated material. The purpose of such in situ methods is to destroy or transform contaminants for purposes of reducing bioavailability and to reduce or remove contaminants from bulk soil (Reed et al. 1992). In situ techniques are favoured over ex situ techniques, because they cost less and have lower ecosystem impact. A conventional ex situ technique is to excavate soil contaminated with heavy metals and remove them for burial at a landfill site (McNeil and Waring 1992; Smith 1993). Such conventional techniques are generally inappropriate, because they merely shift the contamination elsewhere (Smith 1993); moreover, ex situ approaches impose hazards associated with transport of contaminated soil (Williams 1988). Alternatively, dilution of contaminants to a safe level by importing clean soil and mixing it with contaminated soil may be used as an on-site management approach (Musgrove 1991). Plants used in in situ remediation are increasingly important as means to treat selected solid wastes, and some of the key processes and considerations that attend their use are described below.

4.1 Phytoextraction

Plants are capable of absorbing and accumulating metals in their tissues from contaminated soils, sediments and water at high concentrations (Peterson 1975). Such a process is called phytoextraction or phytoaccumulation (U.S.EPA 2000). Plants may constitute the best approach for removing soil contamination, when one wishes to isolate contaminants without destroying soil structure and fertility. Phytoextraction, whether utilized to remove toxic metal or radionuclide contaminants from soils, is best suited for remediation of diffusely polluted areas; such areas have relatively low concentrations of pollutants, and the contaminants occur superficially in soil (Rulkens et al. 1998). Although different approaches have been employed, the two basic phytoextraction strategies that have been used are (i) chelate-assisted phytoextraction or induced phytoextraction, in which artificial chelates are added to treated soil to increase the mobility and uptake of metal contaminants and (ii) continuous phytoextraction, in which the removal of metal depends on the natural physiological ability of the plant. Hyperaccumulator plant species exist that are capable of enhanced removal efficiency and these are the species most employed in continuous phytoextraction. For this technology to be feasible, plants must extract large concentrations of heavy metals into their roots, translocate the heavy metals to surface biomass and produce a large quantity of plant biomass. When phytoextraction is employed, a potentially valuable feature is that the heavy metals taken up by phytoextraction into plant biomass can be captured and recycled (Brooks et al. 1998).

4.2 Phytovolatilization

Phytovolatilization, another phytoremediation process, employs plants that are capable of absorbing contaminants from soil and then transforming them into volatile forms that can be transpired into the atmosphere. Phytovolatilization is a normal process that occurs as trees or other plants grow, absorb and translocate water contaminated with organic and inorganic substances (Bañuelos et al. 1997; Burken and Schnoor 1999). Some contaminants are translocated to leaves and volatilize into the atmosphere, usually at comparatively low concentrations (Mueller et al. 1999; Suszcynsky and Shann 1995; Watanabe 1997). This process has been primarily used for removal of mercury from soil; absorbed mercury is transformed into volatile forms and is transpired into the atmosphere. Moreover, plants transform the mercuric ion into elemental mercury, a less toxic form. Unfortunately, mercury released into the atmosphere by phytovolatilization may be redeposited in the ecosystem through precipitation (Henry 2000).

Some metal contaminants such as As, Hg and Se may naturally exist as gaseous species in the environment. In recent years, researchers have sought naturally occurring or genetically modified plants capable of absorbing elemental forms of these metals from the soil. Once absorbed, plants can biologically convert these metals to gaseous species within the plant and release them into the atmosphere. To date, selenium phytovolatilization has received the most attention in this regard (Bañuelos et al. 1993; Lewis et al. 1966; McGrath 1998; Terry et al. 1992), because this element is a serious problem in many parts of the world where Se-rich soils are prominent (Brooks 1998). According to Brooks (1998), the release of volatile Se compounds from higher plants was first reported by Lewis et al. (1966). In addition, Gary Bañuelos of USDS's Agricultural Research Service has found that some plants grow in high Se media and produce volatile selenium in the form of dimethyl selenide and dimethyl diselenide (Bañuelos 2000). One example, identified as *Astragalus racemosus* was found to emit dimethyl diselenide (Evans et al. 1968). Moreover, selenium was released from alfalfa as dimethyl selenide, though it is not a hyperaccumulator plant for Se. Lewis et al. (1966) showed that both selenium nonaccumulator and accumulator species volatilize selenium. Terry et al. (1992) reported that members of the Brassicaceae are capable of releasing up to 40 g of Se/ha/d as various gaseous compounds. Some aquatic plants, such as cattail (*Typha latifolia* L.), show clear potential for Se phytoremediation (Pilon-Smits et al. 1999).

Unlike other remediation techniques, once contaminants have been removed via volatilization, there is a loss of control over their migration to other areas. Some authors suggest that addition of phytovolatilized contaminants to the atmosphere would not contribute significantly to the atmospheric pollution pool, because the contaminants are probably subject to more effective or rapid natural degradation processes such as photodegradation (Azaizeh et al. 1997). The consequences of releasing metals to the atmosphere must be considered before adopting this method as a remediation tool.

4.3 Rhizofiltration

Rhizofiltration is the process of removing contaminants from flowing water and aqueous waste streams through extensive and massive root uptake by plants. Several aquatic plant species and hyperaccumulator plants have been found to remove heavy metals (Table 4) from waste-water streams. Formally, the definition of rhizofiltration is the use of both terrestrial and aquatic plants to absorb, concentrate and precipitate contaminants from polluted aqueous sources by processing low concentrations of contaminants in their roots. Rhizofiltration can

Table 4 Examples of hyperaccumulator plants

Latin name of the plant	English name	Element/heavy metals	Notes
Brassica juncea L.	Indian mustard	Cd(A), Cr(A), Cu(H), Ni(H), Pb(H), Pb(P), U(A), Zn(H)	Cultivated
Vallisneria americana	Tape grass	Cd(H), Pb(H)	Native to Europe and North Africa; widely cultivated in the aquarium trade
Dicoma niccolifera	–	–	35 documented uses of this plant
Eichhornia crassipes	Water hyacinth	Cd(H), Cu(A), Hg(H), Pb(H), Zn(A)	Pantropical/subtropical. Roots naturally absorb pollutants; some organic compounds believed to be carcinogenic at concentrations 10,000 times that in the surrounding water
Pistia stratiotes	Water lettuce	Cd(T), Hg(H), Cr(H), Cu(T)	–
Salvinia molesta	Kariba weeds or water ferns	Cr(H), Ni(H), Pb(H), Zn(A)	–
Spirodela polyrhiza	Giant duckweed	Cd(H), Ni(H), Pb(H), Zn(A)	Native to North America

H hyperaccumulator, *A* accumulator, *P* precipitator, *T* tolerant
Source: http://en.wikipedia.org/wiki/Phytoremediation,_Hyperaccumulators

be employed to partially treat industrial discharge, agricultural runoff or acidic-mine drainage wastes. Research has shown that rhizofiltration may be effective for removing lead, cadmium, copper, nickel, zinc and chromium, all of which are primarily retained by plant roots (Chaudhry et al. 1998; U.S. EPA 2000). Rhizofiltration has the advantage of being useful for both in situ or ex situ applications and plant species other than hyperaccumulator plant species are effective and can be used (Table 4).

4.4 Phytostabilization

Plants that are metal tolerant may also be employed to reduce the mobility of metals from contaminated sites. The process is called phytostabilization (Salt et al. 1995; Fig. 4). Utilization of phytostabilization processes is sometimes favoured over remediation, because they cost less and require low maintenance (Berti and Cunningham 2000; Cunningham and Berti 1993). Phytostabilization may also be used to remediate mining sites and processing tailings and for revegetating mining areas.

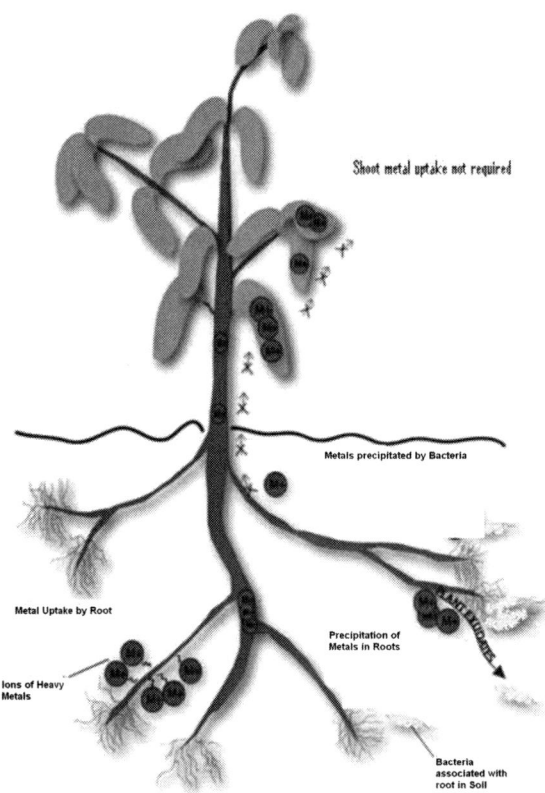

Fig. 4 Schematic picture showing phytostabilization mechanisms (Source: Mendez and Maier 2008)

Several perennial grasses, shrubs and trees (*Quail bush, Anthyllis vulneraria, Festuca arvernensis, Koeleria vallesiana, Armeria arenaria, Lantana camara, Cassia tora, Datura innoxia, B. juncea, Brassica campestris, Phragmites karka, Leersia hexandra*) are being used to revegetate mine-tailing sites. These plants are suitable and effective in achieving phytostabilization. Grasses grow rapidly and provide ground cover that may temporarily limit dispersion of tailings. However, trees and shrubs are important because they provide an extensive canopy and establish a deeper root network that may prevent erosion over the long term. Shrubs or trees provide a environment rich with nutrients for grasses and also reduce moisture stress and improve soil characteristics in arid and semiarid climates (Belsky et al. 1989; Tiedemann and Klemmedson 1973, 2004). Additionally, the establishment of different functional plant species increases plant productivity and yield. Although a few plants may eventually dominate the ecosystem as a result of selection pressure, the presence and effect of less abundant species is still significant in promoting a self-sustainable ecosystem (Tilman et al. 2001). A listing of the different plant species that are being used for phytostabilization is presented in Table 5.

Table 5 Plant families from which potential phytostabilization candidates may be sourced

Plant/family	Metal contaminants	Location	Note
Anacardiaceae			
Pistacia terebinthus Bieberstein	Cu	Cyprus	Field study using 1:1 chicken fertilizer:soil and mine waste
Schinus molle L.	Cd, Cu, Mn, Pb, Zn	Mexico	Plant survey
Asteraceae			
Baccharis neglecta Britt.	As	Mexico	Plant survey
Bidens humilis H.B.K.	Ag, As, Cd, Cu, Pb, Zn	Ecuador	Plant survey
Isocoma veneta (Kunth) Greene	Cd, Cu, Mn, Pb, Zn	Mexico	Plant survey
Viguiera linearis (Cav.) Sch.			
Chenopodiaceae			
Teloxys graveolens (Willd.) W.A. Weber	Cd, Cu, Mn, Pb, Zn	Mexico	Plant survey
Atriplex lentiformis (Torr.) S. Wats.	As, Cu, Mn, Pb, Zn	USA	Greenhouse study using compost
Atriplex canescens (Pursh) Nutt.	As, Hg, Mn, Pb	USA	Field study
Euphorbiaceae			
Euphorbia sp.	Cd, Cu, Mn, Pb, Zn	Mexico	Plant survey
Fabaceae			
Dalea bicolor Humb. & Bonpl. ex Willd.	Cd, Cu, Mn, Pb, Zn	Mexico	Plant survey
Plumbaginaceae			
Lygeum spartum L.	Cu, Pb, Zn	Spain	Plant survey
Poaceae			
Piptatherum miliaceum (L.) Coss.	Cu, Pb, Zn	Spain	Plant survey

Source: Mendez and Maier 2008

4.5 Plants Species Suitable for Phytoremediation

Several plant species can be used to phytoremediate mining and processing tailings and for revegetation of mining sites. Such species are biologically active plants and most are suitable for removal of heavy metal ions. An example of an effective plant species is *B. juncea*. This plant is capable of phytoaccumulating heavy metals from soil to a total content of 897 ppm; such metals are mainly translocated to green leaves. *B. juncea* effectively transports lead from roots to leaves, which is essential for phytoextraction of lead. Another related species, oil rape (*Brassica napus* var. Banacanka), has demonstrated hyperaccumulative capability (Mendez and Maier 2008). This plant may be useful for cleaning the air, ground water, waste water and soil matrices. Research performed with *B. napus, Helianthus annuus, Calamagrostis epigejos, Tussilago farfara, Sisymbrium orientale* has clearly shown that these plants may be useful as phytoremediator species in contaminated terrain.

5 Hyperaccumulation by Plant Species

Some plants accumulate larger amounts of heavy metals in their tissues than do others. A key success factor, when trying to establish an effective phytoremediating plant community, is to find native plant species that grow well in the area to be remediated, but to choose ones that are also effective absorbers of targeted toxic elements from soil. Use of native plants avoids introduction of non-native and potentially invasive new species that could threaten regional plant diversity. Few field trials have yet to take advantage of native plant diversity; not doing so has often resulted in poor plant colonization at waste sites. Some examples of hyperaccumulator plant species are presented in Table 4.

Conesa et al. (2007) recently conducted a greenhouse study to examine metal uptake from tailings by the needlegrass plant *Lygeum spartum*, grown from both seed and rhizomes. Plants grown in the greenhouse from seeds absorbed significantly more metal than did plants grown from rhizomes. However, plants collected from the tailings site itself showed one order of magnitude lower metal accumulation than those tested in the greenhouse. Therefore, one can conclude that prospectively certain entities at the tailings site inhibited uptake into these plants. In fact, an essential point for successful use of a plant in phytostabilization is that it be able to self-propagate successfully, with no additional inputs. The available literature reveals that the long-term fate of metals at revegetated tailings sites has not been explored thoroughly. Such information is needed to evaluate the efficacy of phytostabilization as means to permanently reduce metal toxicity of waste tailing materials.

Different heavy metals behave differently in trees. Pb, Cr and Cu are not very mobile in trees and are retained primarily in roots. In contrast, Cd, Ni and Zn are more easily translocated to the aerial portions of woody plants. Such differences in mobility and storage have important implications for how effective phytoremediation may be as means to control leaching of heavy metals from soils or waste areas.

Two tree species (*Salix viminalis* and *Salix dasyclados*) have considerable potential as vegetative cover for phytoremediation of land contaminated by heavy metals. Evidence from natural establishment of trees on contaminated sites supports the view that some tree types can survive under adverse conditions. Some tree species may not tolerate levels of contamination as high as others, but that does not detract from the utility that these tree species may have for remediation. Such trees may survive because of facultative tolerance, such as avoidance by roots of highly contaminated substrate or by immobilization of heavy metals in the root system. There is no evidence to support a specific, genetically transmitted tolerance system in such plants. However, some evidence exists to show that tolerance may be increased by acclimation of individual trees to low concentrations of heavy metals (Pulford and Watson 2003).

Phytoremediation technology is only in its infancy in India. However, it is a cost-effective and unfolding process that comprises a viable alternative to conventional remedial methods. However, further research results are needed to identify factors that affect what constitutes suitable plant species for remediation and what mine-tailings chemistry is most compatible for utilization of phytoremediation technologies.

6 Summary

Large quantities of iron-ore tailings are being generated annually in the world from mining and processing of iron ores. It has been estimated that around 10–15% of the iron ore mined in India has remained unutilized and discarded as slimes during mining and subsequent processing. Soil contamination resulting from mining activities affects surrounding flora and fauna and presents a large clean-up challenge to the mining industry. Innovative new methodologies have been proposed and among the most promising are those that rely on new phytoremediation technology.

In this paper we address and review the status of phytoremediation as a technology to reduce and control contaminated mine wastes. Several different approaches and different plant species are used to remove environmentally toxic metals from mine waste sites. Such approaches have the objective of restoring mining waste sites to human and animal use, or at least, to curtail or eliminate the off-site movement of toxic entities that potentially could reach humans. How well phytoremediation performs as an alternative soil restoration technology depends on several factors, including the composition of soil, toxicity level of the contaminant, degree to which plant species fit natural local growth patterns and type and concentration of metal/contaminant in such plants.

Phytoremediation has opened prospects for less costly, yet practicable approaches to clean-up contaminated waste sites, particularly those associated with mineral extraction mining. We discuss several plant species that are capable of phytoextracting and/or phytostabilizing harmful elements from contaminated soil and water; such processes are prospectively effective for addressing waste problems that derive from mining and processing activities, as well as those that derive from mitigating the threat posed by waste that surrounds mining sites. Unfortunately, phytoremediation is still in the embryonic stage, and more research is needed to find the plant species that will be most effective for addressing different mining waste scenarios. Such plants must be able to survive and even thrive in heavily contaminated soil and be able to mitigate the pollutants that exist in the soil in which these plants will grow.

References

Alden RW, Butt AJ, Jackman SS, Hall GJ, Young Jr R (1985) Comparison of microcosm and bioassay techniques for estimating ecological effects from open ocean disposal of contaminated dredged sediments. NTIS Report. Old Dominion University, Norfolk, VA, USA

Arnaez J, Larrea V, Ortigosa J, (2004) Surface runoff and soil erosion on unpaved forest roads from rainfall simulation tests in northeastern Spain. Catena 57 1:1–14

Azaizeh HA, Gowthaman S, Terry N (1997) Microbial selenium volatilization in rhizosphere and bulk soils from a constructed wetland. J Environ Qual 26(3): 666–672

Baker AJM, Walker PL (1990) Ecophysiology of metal uptake by tolerant plants. In: Shaw AJ (ed) Heavy metal tolerance in plants: evolutionary aspects. CRC Press, Boca Raton, FL, pp 155–177

Balczon JM, Pratt JR (1994) A comparison of the responses of two microcosm designs to a toxic input of copper. Hydrobiologia 281:101–114

Bandopadhyay A, Kumar R, Ramachandrarao P (eds) (2002) Clean technologies for metallurgical industries. Allied, New Delhi, India

Bandopadhyay A, Kumar S, Das SK, Singh KK (1999) In the pursuit of waste free metallurgy. NML Tech J 41(4):143–162

Bañuelos GS, Cardon G, Mackey B, Ben-Asher J, Wu LP, Beuselinck P (1993) Boron and selenium removal in B-laden soils by four sprinkler irrigated plant species. J Environ Qual 22(4):786–797

Bañuelos GS, Ajwa HA, Mackey LL, Wu C, Cook S, Akohoue S (1997) Evaluation of different plant species used for phytoremediation of high soil selenium. J Environ Qual 26:639–646.

Bañuelos GS (2000) Phytoextraction of selenium from soils irrigated with selenium-laden effluent. Plant and Soil 224(2):251–258

Belsky AJ, Amundson RG, Duxbury JM, Riha SJ, Ali AR, Mwonga SM (1989) The effects of trees on their physical, chemical, and biological environments in a semi-arid Savanna in Kenya. J Appl Ecol 26:1005–1024

Berti WR, Cunningham SD (2000) Phytostabilisation of metals. In: Raskin I (ed) Phytoremediation of toxic metals: using plants to clean up the environment. Wiley-Interscience, New York, NY, pp 71–88

Blaylock MJ, Huang JW (2000) Phytoextraction of metals. In: Raskin I, Ensley BD (eds) Phytoremediation of toxic metals using plants to clean up the environment. Wiley, New York, NY, pp 53–70

Boulet MP, Larocque ACL (1998) A comparative mineralogical and geochemical study of sulfide mine tailings at two sites in New Mexico, USA. Environ Geol 33:130–142

Bradshaw AD, Humphreys MO, Johnson MS (1978) The value of heavy metal tolerance in the revegetation of metalliferous mine wastes. In: Goodman GT, Chadwick MJ (eds) Environmental management of mineral wastes. Sijthoff & Noordhoff, The Netherlands, pp 311–314

Brooks RR (1998) In: Brooks RR (eds) Plants that hyperaccumulate heavy metals Wallingford, CAB International, pp 380–384

Brooks RR, Chambers MF, Nicks LJ, Robinson BH (1998) Phytomining. Trends Plant Sci 1: 359–362

Burken JG, Schnoor JL (1999) Distribution and volatilisation of organic compounds following uptake by hybrid poplar trees. Int J Phytoremediat 1:139–151

Chaudhry TM, Hayes WJ, Khan AG, Khoo CS (1998) Phytoremediation – focusing on accumulator plants that remediate metal contaminated soils. Aust J Ecotoxicol 4:37–51

Conesa HM, Robinson BH, Schullin R, Nowack B (2007) Growth of *Lygeum spartum* in acid mine tailings: response of plants developed from seedlings, rhizomes, and at field conditions. Environ Pollut 145:700–707

Cooper EM, Sims JT, Cunningham SD, Huang JW, Berti WR (1999) Chelate-assisted phytoextraction of lead from contaminated soil. J Environ Qual 28:1709–1719

Cunningham SD, Berti WR (1993) Remediation of contaminated soils with green plants – an overview. In Vitro Cell Dev Biol 29:207–212

Das B, Prakash S, Mohapatra BK, Bhaumik SK, Narasimahan KS (1992) Beneficiation of iron ore slimes using hydrocyclone. Miner Metallurg Process 9(2):101–103

Das B, Ansari MI, Mishra DD (1993) Effective separation of Barsua iron ore slimes using hydrocyclone. Miner Metallurg Process 52:52–55

Das SK, Kumar S, Singh KK (2003) Process for the production of ceramic tiles. Patent no. 13005NF, filed in 2003 in Australia.

Dong J, Wu F, Huang R, Zang G (2007) A chromium-tolerant plant growing in Cr contaminated land. Int J Phytoremediat 9:167–179

Evans CS, Asher C, Johnson CM (1968) Isolation of dimethyl diselenide and other volatile selenium compounds from *Astragalus racemosus* (Pursh.) Aust J Biol Sci 21:13–20

Fichet D, Radenac G, Miramand P (1998) Experimental studies of the impacts of harbour sediments resuspension to marine invertebrate larvae: bioavailability of Cd, Cu, Pb and Zn and toxicity. Mar Pollut Bull 36:509–518

Ghosh M, Singh SP (2005) A review on phytoremediation of heavy metals and utilization of its by-products. Appl Ecol Environ Res 3(1):1–18

Ghosh MK, Sen PK (2001) Characteristics of iron ore tailing slime in India and its test for required pond size. Environ Monitor Assess 68:51–61

Gustavson K, Waengberg SA (1995) Tolerance induction and succession in microalgae communities exposed to copper and atrazine. Aquat Toxicol 32:283–302

Henry JR (2000) An overview of phytoremediation of lead and mercury. National Network of Environmental Management Studies (NNEMS) Fellow, pp 1–31

Huang JW, Chen J, Berti WB, Cunningham SD (1997) Phytoremediation of lead-contaminated soils: role of synthetic chelates in lead phytoextraction. Environ Sci Technol 31:800–805

Hurk PVD, Eertman RHM, Stronkhorst J (1997) Toxicity of harbour canal sediments before dredging and after offshore disposal. Mar Pollut Bull 34:244–249

Johnson MS, Bradshaw AD (1977) Prevention of heavy metal pollution from mine wastes by vegetative stabilisation. Trans Inst Min Metall 86:47–55

Johnson MS, Cooke JA, Stevenson JKW (1992) Revegetation of metalliferous wastes and land after metal mining. In: Hester RE, Harrison RM (eds) Mining and its environmental impact. Royal Society of Chemistry, London, pp 31–47

Kandel D, Western AW, Grayson RB, Turral HN (2004) Process parameterization and temporal scaling in surface runoff and erosion modeling. Hydrol Process 18(8):1423–1446

Krzaklewski W, Pietrzykowski M (2002) Selected physicochemical properties of zinc and lead ore tailings and their biological stabilisation. Water Air Soil Pollut 141:125–142

Kumar R, Kumar S, Mehrotra SP (2005) Fly ash: towards sustainable solutions. In: Proceedings of the international conference fly ash, India, pp 11–12

Kumar S, Singh KK (2004) Effects of fly ash additions on the sintering and physico-mechanical properties of ceramic tiles. J Met Mater Process 16(2–3):351–358

Lewis MA, Weber DE, Stanley RS, Moore JC (2001) Dredging impact on an urbanized Florida bayou: effects on benthos and algal-periphyton. Environ Pollut 115:161–171

Lewis BG, Johnson CM, Delwiche CC (1966) Release of volatile selenium compounds by plants: collection procedures and preliminary observations. J Agric Food Chem 14:638–640

McGrath SP (1998) Phytoextraction for soil remediation. In: Brooks RR (ed) Plants that hyperaccumulate heavy metals: their role in phytoremediation, microbiology, archaeology, mineral exploration and phytomining. CAB International, New York, NY, pp 261–288

McNeil KR, Waring S (1992) In: Rees JF (ed) Contaminated land treatment technologies. Society of Chemical Industry, Elsevier Applied Sciences, London, pp 143–159

Mendez MO, Glenn EP, Maier, RM (2007) Phytostabilization potential of quailbush for mine tailings: growth, metal accumulation, and microbial community changes. J Environ Qual 36:245–253

Mendez MO, Maier RM (2008) Phytostabilization of mine tailings in arid and semiarid environments – an emerging remediation technology. Environ Health Perspect 116(3):278–283

Mueller B, Rock S, Gowswami D, Ensley D (1999) Phytoremediation decision tree – prepared by – interstate technology and regulatory cooperation work group, pp 1–36 http://www.cluin.org/download/partner/phytotree.pdf

Musgrove S (1991) An assessment of the efficiency of remedial treatment of metal polluted soil. In: Proceedings of the international conference on land reclamation, University of Wales. Elsevier Science Publication, Essex, UK

Peterson PJ (1975) Element accumulation by plants and their tolerance of toxic mineral soils. In: Hutchinson TC (ed) Proceedings of the International Conference on Heavy Metals in the Environment. University of Toronto, Canada, 2:39–54

Pilon-Smits EAH, Desouza MP, Hong G, Amini A, Bravo RC, Payabyab ST, (1999) Selenium volatilization and accumulation by twenty aquatic plant species. J Environ Qual 28(3): 1011–1017

Pulford ID, Watson C (2003) Phytoremediation of heavy metal-contaminated land by trees – a review. Env Internat 29:529–540.

Reed D, Tasker IR, Cunnane JC, Vandegrift GF (1992) Environmental restoration and separation science. In: Vandgrift GF, Reed DT, Tasker IR (eds) Environmental Remediation Removing Organic and Metal Ion Pollutants. ACS Symposium Series 509 Amer Chem Soc, Washington DC, pp 1–21

Rulkens WH, Tichy R, Grotenhuis JTC (1998) Remediation of polluted soil and sediment: perspectives and failures. Water Sci Technol 37:27–35

Salt DE, Blaylock M, Kumar PBAN, Dushenkov V, Ensley BD, Chet I, Raskin I (1995) Phytoremediation: a novel strategy for the removal of toxic metals from the environment using plants. Biotechnology 13: 468–474

Smith B (1993) Remediation update funding the remedy. Waste Manage Environ 4:24–30

Southam G, Beveridge TJ (1992) Enumeration of *Thiobacilli* within pH-neutral and acidic mine tailings and their role in the development of secondary mineral soil. Appl Environ Microbiol 58:1904–1912

Suszcynsky EM, Shann JR (1995) Phytotoxicity and accumulation of mercury subjected to different exposure routes. Environ Toxicol Chem 14:61–67

Tiedemann AR, Klemmedson JO (1973) Nutrient availability in desert grassland soils under mesquite (*Prosopis juliflora*) trees and adjacent open areas. Proc Soil Sci Soc Am 37: 107–111

Tedemann AR, Klemmedson JO (2004) Responses of desert grassland vegetation to mesquite removal and regrowth. J Range Manage 57:455–465

Terry N, Carlson C, Raab TK, Zayed A (1992) Rates of selenium volatilization among crop species. J Environ Qual 21:341–344

Tilman D, Reich PB, Knops J, Wedin D, Mielke T, Lehman C (2001) Diversity and productivity in a long-term grassland experiment. Science 294:843–845

Togna MT, Kazumi J, Sabine A, Kirtay V, Young LY (2001) Effect of sediment toxicity on anaerobic microbial metabolism. Environ Toxicol Chem 20:2406–2410

U.S. EPA (Environmental Protection Agency Reports) (2000) Introduction to phytoremediation. EPA 600/R-99/107. National Risk Management Research Laboratory, Cincinnati, OH. http://www.epa.gov/swertio1/download/remed/introphyto.pdf

U.S. EPA (2003) EPA draft report on the environment. June 2003. EPA document no. EPA-260-R-02-006

Walder IF, Chavez WX (1995) Mineralogical and geochemical behavior of mill tailing material produced from lead–zinc skarn mineralization, Hanover, Grant County, New Mexico, USA. Environ Geol 26:1–18

Watanabe ME (1997) Phyto-remediation on the brink of commercialization. Environ Sci Technol 31:182–186

Williams GM (1988) Integrated studies into groundwater pollution by hazardous waste. In: Gronow JR, Schofield AN, Jain RK (eds) Land Disposal of Hazardous Waste: Eng Environ Issues Chichester, UK: Ellis Horwood. 8:37–48

Word JQ, Hardy JT, Crecelius EA, Kiesser SL (1987) A laboratory study of the accumulation and toxicity of contaminants at the sea surface sediments proposed for dredging. Mar Environ Res 23:325–338

Wong JWC, Ip CM, Wong MH (1998) Acid-forming capacity of lead–zinc mine tailings and its implications for mine rehabilitation. Environ Geochem Health 20:149–155

Ye ZH, Shu WS, Zhang ZQ, Lan CY, Wong MH (2002) Evaluation of major constraints to revegetation of lead/zinc mine tailings using bioassay techniques. Chemosphere 47:1103–1111

Fugitive Dust and Human Exposure to Heavy Metals Around the Red Dog Mine

Elizabeth J. Kerin and Hsing K. Lin

Contents

1	Introduction	49
2	Heavy Metal Toxicology	51
3	Dust-borne Metal Exposure Pathways Around Red Dog Mine	53
	3.1 Direct Inhalation	53
	3.2 Contact with Contaminated Soils	54
	3.3 Residential Exposure to Mine Dust	55
	3.4 Subsistence Land Use and Wildlife Effects	56
4	Epidemiological Studies Around Red Dog Mine	57
	4.1 Environmental Exposure Studies	57
	4.2 Occupational Exposure Studies	58
5	Regulatory Oversight of Dust and Metals at Red Dog Mine	60
6	Dust Control Measures Around Red Dog Mine	60
7	Summary	61
References		62

1 Introduction

The Red Dog Mine is a high-grade open pit lead–zinc mine located in the northwestern Brooks Range, about 130 km north of Kotzebue, Alaska (Kral 1992) (Fig. 1). The mine began operation in 1989 and exploration has revealed deposits such as the Aqqaluk Deposit that would allow mining to continue until 2031 (Liles 2006; USEPA 2007). The mine is operated at a rate of 5.4 kt/d (6,000 short t/d). The mined ore is processed through crushing and grinding circuits, followed by froth flotation separation to produce zinc and lead concentrates in the mineralogical forms of sphalerite (zinc sulfide) and galena (lead sulfide), respectively. Because

E.J. Kerin (✉)
University of Alaska Fairbanks, Fairbanks, AK, USA
e-mail: lizkerin@hotmail.com

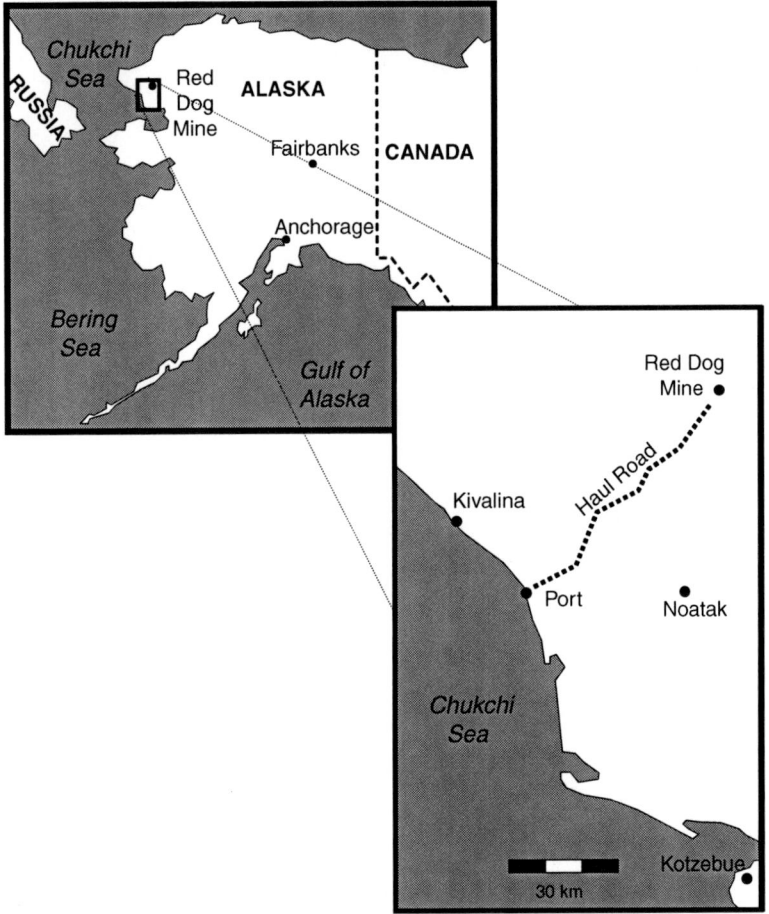

Fig. 1 Map showing approximate location of Red Dog Mine (68°4′19″N 162°52′34″W), DeLong Mountain Regional Transportation System port, and haul road (distances are approximate)

the sphalerite and galena are fine-grained, the ore is finely ground for effective separations in the flotation circuit. As a result, both the zinc and lead concentrates are very fine in particle size, with 80% of the final concentrates passing 30 and 20 μm screens, respectively. The flotation concentrates are dewatered using pressure filters, and the resultant dehydrated concentrates contain 7.5–8.0% moisture. The annual production of zinc and lead concentrates are 508 and 109 kt, respectively (Kral 1992). Dewatered zinc and lead concentrates produced at the Red Dog Mine site are hauled a distance of 84 km by truck along an unpaved road to the DeLong Mountain Regional Transportation System port on the Chukchi Sea, where they are then shipped to international markets (Liles 2006).

Creation of fugitive dust by haul activities has been observed along the DeLong Mountain Transportation System road (Ford and Hasselbach 2001). Fugitive dust consists of non-point re-suspended particulate matter and is of health concern along the unpaved haul roads found in and near industrial sites such as mines, quarries, and asphalt plants (USEPA 1998). At the Red Dog Mine, fugitive dust having elevated metal concentrations constitutes a source of human exposure to lead and other metals (Exponent Engineering and Scientific Consulting 2008). Moreover, the area around the Red Dog Mine and the haul road is used by Native Alaskans for subsistence activities including hunting, berry picking, and fishing; and winter trails that connect area villages cross the haul road at multiple locations. Dust and fugitive emissions along the haul road present a potentially important pathway of metal exposure to wildlife and human subsistence communities located in the area. The potential for occupational exposure to dust-borne lead also presents a health concern to Red Dog Mine workers. In this chapter, we address the environmental impact of dust-borne heavy metal contamination in the vicinity of the Red Dog Mine, its potential effect on traditional land use around the mine, port, and haul road, and control technologies that can be used to mitigate this impact.

2 Heavy Metal Toxicology

The deleterious human health effects of heavy metals are well documented (Klaassen et al. 1996). The content of the lead concentrate transported along the Red Dog Mine haul road is approximately 60% lead, 0.1% cadmium, and 10% zinc; zinc concentrate contains about 55% zinc, 3% lead, and 0.3% cadmium (Exponent Engineering and Scientific Consulting 2007). Of these metals, overexposure to zinc is of the least concern to human populations, because zinc concentration is well regulated in the human body, and excessive levels require heavy exposure (Klaassen et al. 1996). Most humans contain about 2–3 g of zinc, with mean daily dietary zinc intakes ranging from 4.7 to 18.6 mg/d (Maret and Sandstead 2006). It would be highly unlikely for an individual to exceed these levels through environmental exposure, even in an environment impacted by zinc mining. Concerns of adverse effects from zinc exposure are generally focused on overuse of zinc dietary supplements (Maret and Sandstead 2006) or serious health effects resulting from zinc deficiency (Klaassen et al. 1996).

Cadmium exposure is of concern at the Red Dog Mine, but does not pose a human health risk as great as that of lead exposure. Renal toxicity is a recognized human health effect of cadmium. Study results indicate that cadmium intake by ingestion should be kept below 30 μg/person/d to avoid renal impairment (Satarug et al. 2000). Endocrine disruption and detrimental effects on mammalian reproduction have also been documented as a result of low-level cadmium exposure, but these effects have yet to be completely elucidated (Henson and Chedrese 2004). Inhalation exposure pathways of cadmium are also important, because most airborne cadmium is respirable, and 15–30% of inhaled cadmium can be absorbed by the lungs (Klaassen et al. 1996). The toxic effect most associated with chronic inhalation exposure to cadmium is chronic obstructive pulmonary disease (Klaassen et al. 1996). The average cadmium concentration in the Red Dog zinc concentrate is 0.3%. The US Environmental Protection Agency (USEPA) National Ambient Air Quality Standard has a 24-hr average concentration limit of 150 $\mu g/m^3$. Assuming that the concentration of particulate matter smaller than 10 μm in the Red Dog Mine area is equal to the USEPA standard, cadmium air concentrations would be about one order of magnitude below the 5 $\mu g/m^3$ 8 hr Occupational Safety and Health Administration permissible exposure limit (PEL; Code of Federal Regulations Title 29 Part 1910.1027, 2007). The PEL for chronic health effects of inhaled cadmium appears to be protective of the potential exposure level that exists in the Red Dog Mine area.

Lead poses the greatest human health risk at the Red Dog Mine, because of its toxicity and the high concentrations of lead in the ore. Neurotoxic effects of lead are recognized in both children and adults (Fig. 2). In children, central nervous system impairment can occur when blood lead levels approach 10 $\mu g/dL$. Although adverse effects have been observed below 10 $\mu g/dL$ (Canfield et al. 2003), this concentration is recognized by the Centers for Disease Control and Prevention (CDC) as the threshold level of concern for child blood lead levels. Impairment at these levels manifests themselves as cognitive, growth, and behavioral problems and do

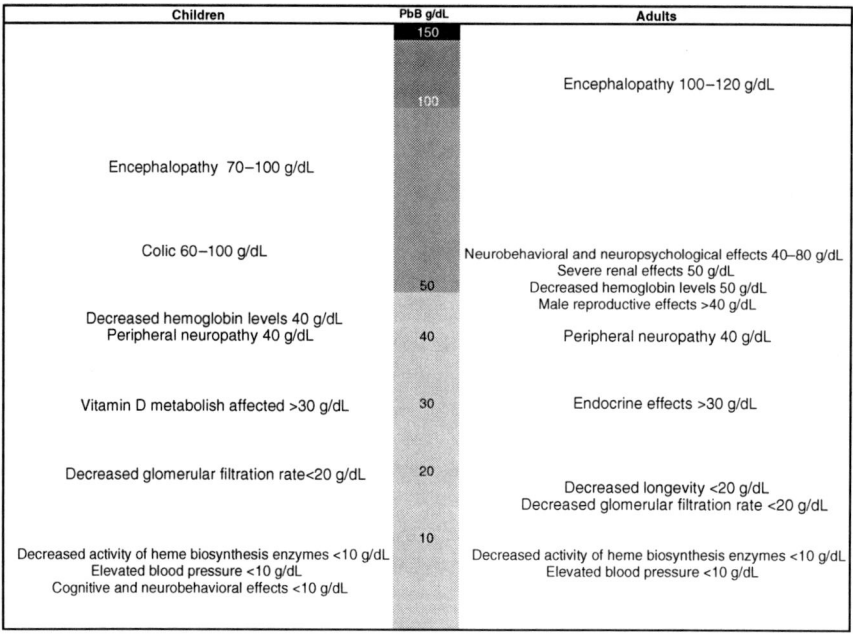

Fig. 2 Effects of inorganic lead on children and adults – lowest observable adverse health effects. Figure adapted from Agency for Toxic Substances and Disease Registry (1992) using data from Agency for Toxic Substances and Disease Registry (2007)

not seem to be reversible even as children age and their exposure and sensitivity to lead are reduced (Murgueytio et al. 1998). Neurotoxic effects of lead are also recognized in adults at blood lead levels exceeding ~25 µg/dL. In adults, lead impairment affects the peripheral nervous system and appears to be more reversible than in children (Bellinger 2004). Long-term exposure to lead has also been noted as a risk factor for the development of hypertension in adults (Hu et al. 1996).

3 Dust-borne Metal Exposure Pathways Around Red Dog Mine

Pathways of human exposure to dust-borne heavy metals around Red Dog Mine include inhalation, contact with contaminated soils and dust, occupational exposure, and hunting or harvesting of plants and animals exposed to dust (Exponent Engineering and Scientific Consulting 2007).

3.1 Direct Inhalation

Dust particles with diameters >10 µm are trapped in the mid-respiratory tract, and can be coughed up and ingested, resulting in metal absorption by the gastrointestinal tract. Smaller diameter particles are carried into the alveoli, where metals may be absorbed into the bloodstream through the alveolar epithelium (Foulkes 1998). Potential dust exposure sources along the haul road include spilled ore concentrate, materials tracked out of the mine on trucks, and re-suspension of road dust.

Humans may be at risk for direct inhalation of dust at the Red Dog Mine by being in proximity to the road when dust is generated or through widespread particulate pollution generated by mining activities. There is potential for humans to be sufficiently near the haul road to be exposed to dust produced by truck traffic. Although the road does not pass near any residences, Kivalina and Noatak residents use the area around the road for subsistence activities (Dames & Moore, Inc. Consulting 1983); moreover several winter trails cross the road (Exponent Engineering and Scientific Consulting 2007). Practices used to limit human exposure to dust include restriction of access to the road, communication of exposure risk to those who use nearby land and trails, and regulatory limits on dust production. Red Dog Mine's Alaska Department of Environmental Conservation Air Quality permit requires the facility to limit public access to the land within 100 m of the haul road. The facility is also required to post signs at winter trail crossings to warn the public to avoid the road when traffic is passing (Alaska Department of Environmental Conservation 2003). The USEPA National Ambient Air Quality Standards limit the 24-hr average concentration of dust emitted from facilities to 150 µg/m^3 (USEPA 2004). Modeling indicates that these limits are not exceeded at the boundaries of the mine, port, or haul road corridor (Exponent Engineering and Scientific Consulting 2007).

There is also potential exposure to levels of elevated particulate matter on a regional scale around the Red Dog Mine. To determine the level of such exposure, air monitoring was conducted for particulates in the nearby villages of Kivalina, 25 km north of the port, and Noatak, 65 km south of the mine. Results showed that average quarterly air concentrations were two orders of magnitude lower than the National Ambient Air Quality Standards (Exponent Engineering and Scientific Consulting 2007). This suggests that human exposure to ambient regional particulate metal pollution is not of concern.

3.2 Contact with Contaminated Soils

The relationship between blood lead levels and exposure to lead in soils and household dust around mining areas has been investigated in many studies. For example, one study in the United Kingdom identified significantly higher concentrations of lead in soils and household dust, in a historical mining area, compared with other areas of Britain, Scotland, and Wales, where mining did not occur (Thornton et al. 1990). The authors of this study also identified a statistically significant relationship between lead levels in household dust and blood lead levels in 2-year-old children. Another study, performed in the Old Mining Belt of southeast Missouri, disclosed elevated blood lead levels in 17% of children living in mining areas, compared with 3% in children living outside of mining areas. Soil and dust levels in mine areas were 10 times higher than in non-mining areas (Murgueytio et al. 1998).

Because of the remote location of Red Dog Mine, exposure to contaminated soil is unlikely to be a concern in residential areas. The air-borne lead concentration measured at Noatak and Kivalina is not high enough to suggest that deposition of particulate matter on soils would result in soil lead concentrations of health concern to residents in these nearby settlements. Because workers do not travel between the mine and residences by the road system, there is no opportunity for vehicles to track ore concentrate or contaminated soil from the mine site to workers' homes.

Human exposure to contaminated soils is of greater concern along roadsides and around the port. Thirty-one lead and zinc concentrate spills, ranging between 1 and 72 metric tons each, occurred along the haul road between 1990 and 2007 (Exponent Engineering and Scientific Consulting 2007). Spilled concentrate was cleaned up as soon as possible, as permitted by the prevailing weather conditions (Exponent Engineering and Scientific Consulting 2007). In addition, small amounts of concentrate have occasionally escaped from haul trucks, and mine materials have been tracked onto the haul road by truck tires or have fallen from trucks onto the roadway. A study of the metal content of mud from wheel-wells of Red Dog vehicles, and in haul road soil samples, showed that both were enriched in cadmium, zinc, and lead when compared with a reference mud sample collected from a vehicle in Kotzebue, Alaska (Brumbaugh and May 2008).

Soil metal concentrations have been measured in the area around the Red Dog Mine, the haul road, and the port. Ford and Hasselbach (2001) analyzed soil and moss samples for heavy metals (*Hylocomium splendens*) along six transects of the haul road, with each transect extending 1,800 m to the north and south of the road.

H. splendens is a non-vascular plant that collects most of its nutrition from precipitation and air-borne particles. Therefore, this moss can be used to estimate the contribution of metals arriving in a certain area from adherence to air-borne dust, rather than solely representing natural soil concentrations (Ford and Hasselbach 2001). Along each transect, lead, zinc, and cadmium concentrations in moss were highest directly along the road and decreased exponentially with distance from the road. Lead concentrations were greater than 60 mg/kg in all transect points within 100 m of the road. These values were 4–7 times higher than the concentration of lead in *H. splendens* collected along the Dalton Highway (a 666-km gravel supply road for the Trans-Alaska Pipeline System and North Slope oil industries) and were comparable to the maximum concentrations reported in polluted areas of central Europe. In addition, Hasselbach et al. (2005) investigated metal concentrations in soil along transects of the haul road. In subsurface soils, metal concentrations did not change with proximity to the road, indicating that dust measured in moss was from a depositional source on the surface of the road and surrounding land.

A study of lead in dust-bearing soil along the haul road and at the port showed surface concentrations of 5,000 mg/kg at both locations (Kelley and Hudson 2007). At depths of 10–30 cm, lead concentrations in soils were 2–3 orders of magnitude less than in surface soils (Fig. 3). This result supports the work of Hasselbach and others and suggests a surface source of metal contamination along the haul road and within the port.

3.3 Residential Exposure to Mine Dust

In addition to the environmental exposures described above, another potential exposure pathway to metals from mining dust at Red Dog Mine is via mine workers

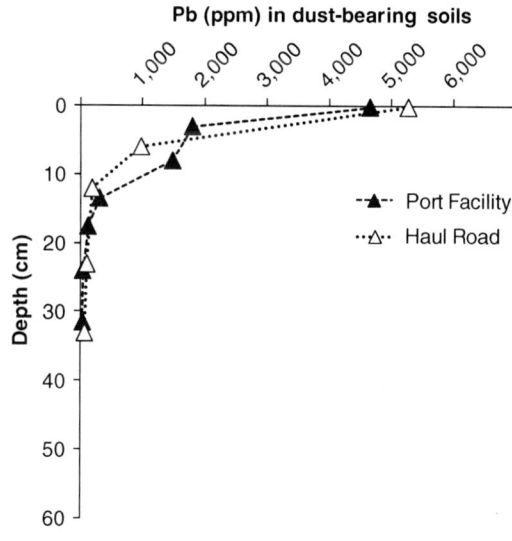

Fig. 3 Depth profile of lead in dust-bearing soils immediately adjacent to the haul road and port facility (from Kelley and Hudson 2007, Fig. 2b, used with permission from AAG/Geological Society of London)

returning home with dust contamination on themselves and their belongings. Studies elsewhere have shown the importance of this style of exposure. For example, lead isotope data taken from households of employees of a lead–zinc–copper mine in New South Wales demonstrated this potential pathway for lead contamination (Chiaradia et al. 1997). Between 40 and 100% of the lead in dust from miners' households were considered to be mine-derived. Although children in these households did not have blood lead levels >10 µg/dL, 20% of their blood lead was from mining lead. This indicates that although children were exposed to multiple lead sources – including leaded paint and leaded gasoline – lead mining comprised an important exposure pathway (Chiaradia et al. 1997).

The Red Dog Mine's remote location probably limits exposure via contamination transported by or on employees. The mine requires workers to use protective gear and to follow hygiene guidelines designed to minimize lead exposure (Bluemink 2008). Almost all workers travel home from the mine by aircraft, and depending on work shifts and plane schedules, workers may shower, wash laundry, and change clothes before leaving the mine, thus limiting the exposure to mine dust in their households.

3.4 Subsistence Land Use and Wildlife Effects

Subsistence activities are a potential source of human exposure to dust-borne heavy metals from mines. Lands near the Red Dog Mine, haul road, and port are used for subsistence hunting, fishing, and gathering by residents of nearby villages and towns (Dames & Moore, Inc. Consulting 1983). One-third of households in this area were dependent on subsistence food at the time of a 1984 study (USEPA 1984). The continued viability of subsistence food is integral to the survival of traditional cultures' economic and nutritional health, as well as spiritual values.

Brumbaugh and May (2008) examined potential sublethal effects of metal exposure in voles and small birds in the vicinity of the Red Dog haul road. Concentrations of lead, aluminum, barium, zinc, and cadmium were measured in the liver and blood from animals captured along the haul road, near the port site, and at a reference site about 96 km south of the port. Blood and liver lead concentrations in animals captured near the haul road and port were 20 times higher than that in animals from the reference site, whereas concentrations of barium, zinc, and aluminum were not comparatively higher than the reference samples. Notwithstanding, histological analysis did not reveal DNA damage or lesions associated with metal poisoning. These results indicate that, although mine lead has impacted lead concentrations within specific terrestrial organisms around the mine site, it does not appear to have had a measurable negative impact on the health of the organisms.

Potential health effects of metals on animals at higher trophic levels were investigated in a study by O'Hara et al. (2003). Concentrations of metals and body condition were studied in caribou from a mass mortality event in western Alaska and in caribou killed by hunters around the Red Dog Mine and around two reference sites. Although caribou in the area of the Red Dog Mine had elevated levels of lead in their liver, kidney, and rumen tissues when compared with animals harvested at the reference sites, concentrations were an order of magnitude below observed toxic levels in toxicological studies of cattle in the continental USA. Analysis of body condition indicated that caribou found dead had died from emaciation rather than from metal toxicosis. The authors of this study did not investigate the source of lead found in animal tissues or the potential human exposure to lead from eating caribou killed in the vicinity of the Red Dog Mine.

The Alaska Department of Health and Social Services has investigated the potential for human exposure to metals through subsistence. Heavy metals have been measured in Dolly Varden salmon (*Salvelinus malma*), Arctic grayling (*Thymallus arcticus*), Arctic caribou (*Rangifer tarandus*), and washed and unwashed salmonberries, in the vicinity of the Red Dog Mine. Estimated ingestion rates of each food source were used, along with USEPA oral reference doses for heavy metals, to determine if humans were exposed to harmful levels of heavy metals through subsistence. The Alaska Department of Health and Social Services concluded that concentrations of heavy metals in each of the above food types do not pose public health concerns, either for children or for adults. In particular, using a USEPA Uptake Model, it was predicted that lead exposure through subsistence would result in blood lead levels below the 10 μg/dL level of concern for children set by the CDC (Alaska Department of Health and Social Services 2001).

4 Epidemiological Studies Around Red Dog Mine

4.1 Environmental Exposure Studies

A number of epidemiological studies have been conducted in villages surrounding the Red Dog Mine (Alaska Department of Health and Social Services 2001). In these studies, blood lead levels were measured in 90% of Kivalina residents and 91% of Noatak residents in 1990; in 21 Medicaid-eligible children in Kivalina in 1993; and in six children in Point Hope in 1992. Across all resident age groups and all studies, average blood lead levels were less than the 10 μg/dL CDC level of concern for children. In Kivalina, 1 child out of 125 tested had a blood lead level >10 μg/dL. Seven out of 128 children in Noatak had blood lead levels >10 μg/dL; of these, only two children had blood lead levels >15 μg/dL (i.e., 20 and 21 μg/dL, respectively). The Alaska Department of Health and Social Services reported that no source of exposure was found that accounted for the higher blood lead levels in these children (Alaska Department of Health and Social Services 2001).

The results of the epidemiological studies in villages surrounding the Red Dog Mine were also compared to blood lead level results measured in residents of Skagway in 1989 and to the CDC National Health and Nutrition Examination

Survey (NHANES) results conducted between 1991 and 1994 (Pirkle et al. 1998). In Skagway, soil concentrations of lead are elevated because of the transportation of lead ore through that city between 1969 and 1982 in railroad cars from a mine in the Yukon Territory (Alaska Department of Health and Social Services 1989). Average blood lead levels of Skagway residents were higher than those of residents of Noatak, Point Hope, and Kivalina. However, all Skagway residents except ore-terminal workers had blood lead levels of 20 μg/dL or less, and the average blood lead level was <10 μg/dL. The results of the Skagway study show that blood lead levels may be below levels of public health concern, even in an environment impacted by lead contamination, and that Red Dog area residents are less affected than those in Skagway (Alaska Department of Health and Social Services 2001).

A comparison of these exposure results with those of the NHANES study also indicates that human health is being protected around the Red Dog Mine. The average blood lead level in children between the ages of 1 and 5 was 2.7 μg/dL in the NHANES study group; in 1990, the average was 2.5 and 3.7 μg/dL in Kivalina and Noatak, respectively (Alaska Department of Health and Social Services 2001). These data show that Red Dog area residents generally do not have blood lead levels of concern from their environmental lead exposure, although it is worth noting that the differences between these blood lead levels were not statistically analyzed. Also not performed were isotopic studies used to elucidate the environmental source of blood lead in children in the vicinity of the Red Dog Mine.

4.2 Occupational Exposure Studies

Occupational exposure to dust-borne lead has the potential to impact the health of workers at the Red Dog Mine. A State of Alaska Epidemiology Bulletin (Wenzel 2008) summarized adult occupational exposure to lead within the State of Alaska. The occurrence of blood lead levels >25 μg/dL was more frequent in Alaska than in the continental USA, between 1996 and 2005. Of adults tested in Alaska, about 11% have blood lead levels exceeding 25 μg/dL and about 1% have blood lead levels exceeding 40 μg/dL. Of the adults with blood lead levels >25 μg/dL, 94% worked in the mining industry. This reflects the significance of mining as a potential occupational pathway of lead exposure, as well as the predominance of the mining industry throughout Alaska and the comparatively low number of manufacturing or construction jobs available.

The Red Dog Mine follows or exceeds US Occupational Safety and Health Administration regulations for the monitoring and medical management of workers' lead exposure. These regulations require annual testing of worker blood lead levels, with increased monitoring if blood lead levels >40 μg/dL. When blood lead concentrations exceed 50 μg/dL, the worker is removed from lead exposure (Kosnett et al. 2007). The Red Dog Mine tests workers at high risk of occupational exposure twice annually and increases the frequency of testing if blood lead levels exceed 25 μg/dL (Bluemink 2008).

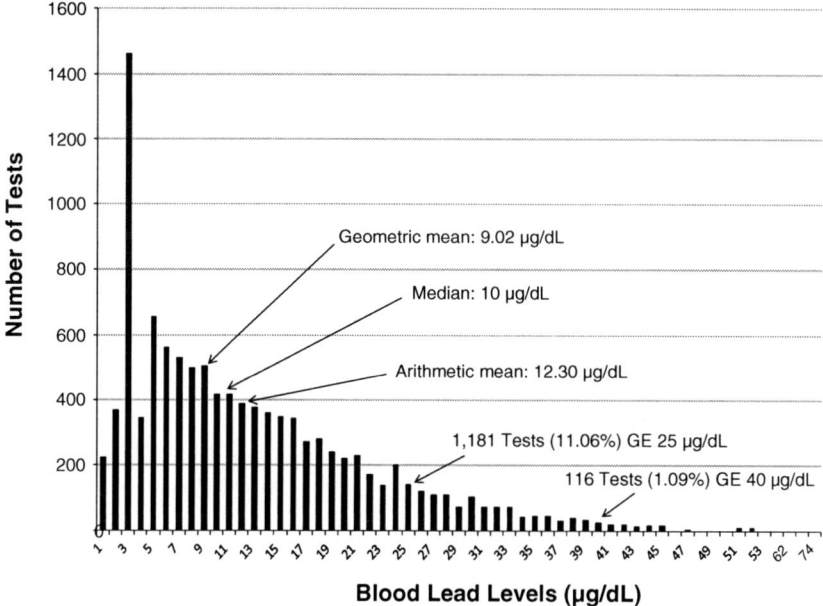

Fig. 4 Red Dog Mine blood lead levels, representing 10,685 samples taken from 1,805 employees. The abbreviation "GE" stands for "greater than or equal to." Figure reprinted with permission from Alaska Department of Health and Social Services (2001)

In 2001, Teck Cominco Ltd, the mining company operating the Red Dog Mine, provided blood lead level data from tests on Red Dog Mine employees and contractors performed between 1992 and 2001 (Alaska Department of Health and Social Services 2001) (Fig. 4). The total dataset included 10,685 individual samples taken from 1,805 employees or contractors. The median blood lead level was 10 μg/dL, with 1,181 (11%) tests >25 μg/dL and 116 (1%) > 40 μg/dL. Presumably, given Teck Cominco's testing protocol, at least some of the tests that exceed 25 and 40 μg/dL represent repeated tests of a subset of individuals with elevated blood lead levels. Red Dog's senior environmental coordinator has reported that only two workers were removed from their jobs since 1989, because of high lead exposure (Bluemink 2008).

The median blood lead level in the Red Dog tests and the occurrence of tests that exceed 25 and 40 μg/dL are comparable to the average blood lead levels of Alaskan workers and the percentage of Alaskan workers with blood lead levels > 25 and 40 μg/dL (Wenzel 2008). Nationally, blood lead levels averaged < 3 μg/dL, in 2001, lower than the average blood lead level suggested by the 10 μg/dL median blood lead level result of the Red Dog tests. In a study of workers in 21 continental states between 1998 and 2001, blood lead levels > 25 and 40 μg/dL occurred 13.4 and 3.9 times per 100,000 workers, respectively (Roscoe et al. 2002). In Alaska, blood lead levels > 25 and 40 μg/dL occurred 14.9 and 1.4 times per 100,000 workers between the years of 1995 and 2006, respectively (Wenzel 2008).

5 Regulatory Oversight of Dust and Metals at Red Dog Mine

Regulatory oversight of the Red Dog Mine has been implemented to ensure that human and environmental health is protected. In 2005, Teck Cominco, Inc. and the Alaska Department of Environmental Conservation entered into a Memorandum of Understanding regarding fugitive dust at the Red Dog Mine (Alaska Department of Environmental Conservation 2005); the memorandum was reinstated and amended in 2007. In this memorandum, Teck Cominco agreed to evaluate human health risks associated with fugitive dust (Alaska Department of Environmental Conservation 2007). As a result, Exponent Engineering and Scientific Consulting completed a fugitive dust risk analysis in 2007. The consulting company then used the results of the study to develop the mine's fugitive dust risk management plan. A draft of this document was completed in August of 2008 (Exponent Engineering and Scientific Consulting 2008).

The Alaska Department of Health and Social Services has worked in collaboration with villages, Maniilaq Health Corporation, and the Northwest Arctic Native Association to conduct epidemiological studies in the area of the Red Dog Mine (Alaska Department of Health and Social Services 2001). Given the length of the mine's potential operating period, the Alaska Department of Health and Social Services has recommended that a formal process for monitoring public health in the area should be developed, and that state regulatory agencies should play a role in assessing data collected by this monitoring plan.

The Alaska Department of Natural Resources provides oversight of mining operations, particularly when reclamation and closure plans or supplemental environmental impact statements are submitted. At Red Dog, scoping documents and a draft environmental impact statement for the development of the Aqqaluk Deposit at Red Dog have been submitted (Tetra Tech 2008). Measures to control fugitive dust and metal contamination will be scrutinized as these documents are reviewed by the Alaska Department of Natural Resources.

6 Dust Control Measures Around Red Dog Mine

Dust control measures are used at the Red Dog Mine and along the haul road to reduce potential environmental and human health effects of dust and metals around the mine. The most common method of dust control along haul roads is the regular application of water to dirt and gravel roads. Watering roads hourly within a mine has been shown to reduce total suspended particles around the mine by 40%. Chemicals may also be applied to control dust along haul roads. Magnesium chloride applications have been shown to reduce dust, generated by haul trucks, by 95% (Reed and Organiscak 2007). The Red Dog Mine waters roads within the mine area itself, and calcium chloride is applied along the haul road (Exponent Engineering and Scientific Consulting 2008).

In addition to watering and chemical treatment, the mine began using improved haul trucks in 2001 to limit spills of concentrate ore. The new trucks have solid sides and steel covers that hydraulically close to prevent the escape of concentrate during shipping. The new trucks are designed to reduce rollover, thereby preventing accidents and potential concentrate spills. The trucks are also designed to produce less dust during unloading, thus limiting the amount of dust created at the port site (Exponent Engineering and Scientific Consulting 2008).

Methods have been implemented to control the amount of dust tracked out of the mine area by haul trucks. The concentrate storage building at the mine was modified to prevent trucks from driving over concentrate during the loading process. Gratings were installed in the area where trucks are loaded, so that concentrate does not fall to the floor where it can be picked up by truck tires. In summer months, a truck-washing station is used to remove dust from the outsides of haul trucks before they leave the concentrate storage area (Exponent Engineering and Scientific Consulting 2008).

A potential method for removing the source of fugitive dust along the haul road is to construct a pipeline that carries concentrate, in slurry form, from the mine to the port. In a draft supplemental environmental impact statement outlining potential scenarios for development of the Aqqaluk Deposit, construction of a concentrate pipeline was included as a potential alternative to trucking concentrate (Tetra Tech 2008). This technology has been successfully applied in a number of mines, including the Antamina Mine, which was brought into production by Teck Cominco, Ltd and Noranda, Inc. Antamina Mine is a copper–zinc mine located approximately 300 km from the coast of Peru. Concentrate slurry is shipped in an underground pipeline to the Punta Lobitos Port, where it is then shipped to market (Compañia Minera Antamina S.A. 2006). Development of such a pipeline at the Red Dog Mine would require specialized arctic engineering techniques, but would reduce the potential for concentrate spills as well as the amount of traffic along the current haul road.

7 Summary

Fugitive dust from the Red Dog Mine is a potential source of exposure to heavy metals for residents of the surrounding area. Possible pathways of exposure include direct inhalation of particles, dermal contact with or ingestion of contaminated soils, residential exposure of individuals who have close association with mine workers, and subsistence activities. Study results indicate that soils and mosses close to the haul road are contaminated with dust and metals from hauling activities. However, investigations of exposure from subsistence activities performed near Red Dog Mine do not indicate that human health has been negatively affected by metal contamination. Epidemiological studies of nearby village residents do not show blood lead levels that exceed the CDC level of concern for children. The mine currently uses several control practices to reduce dust and control human dust and metal exposure. Nonetheless, the potential for human health impairment will persist throughout the life of the mine and beyond. Sound environmental management and monitoring of human health should remain a priority for the Red Dog Mine and for agencies that provide regulatory oversight to the mine.

Acknowledgments The authors thank Christopher H. Conaway for his helpful review of this manuscript.

References

Agency for Toxic Substances and Disease Registry (1992) Case studies in environmental medicine: lead toxicity. U.S. Department of Health and Human Services, Public Health Services, Atlanta, GA

Agency for Toxic Substances and Disease Registry (2007) Toxicological profile for lead. U.S. Department of Health and Human Services, Public Health Services, Atlanta, GA

Alaska Department of Environmental Conservation, Division of Air Quality (2003) Red Dog Mine Facility Title V Operating Permit. Issued to: Teck Cominco Alaska, Inc. Permit No. AQ0290TVP01

Alaska Department of Environmental Conservation, Division of Air Quality and Teck Cominco Alaska, Inc. (2005) Memorandum of understanding between the state of Alaska Department of Environmental Conservation and Teck Cominco Alaska, Inc. relating to the fugitive dust at Red Dog Mine. http://www.dec.state.ak.us/air/reddog.htm. Accessed May 09, 2010

Alaska Department of Environmental Conservation, Division of Air Quality and Teck Cominco Alaska, Inc. (2007) Memorandum of understanding between the state of Alaska Department of Environmental Conservation and Teck Cominco Alaska, Inc. relating to the fugitive dust at Red Dog Mine (Restated and amended). http://www.dec.state.ak.us/air/reddog.htm. Accessed May 09, 2010

Alaska Department of Health and Social Services, Division of Public Health (1989) Health hazard and risk assessment from exposure to heavy metals in ore in Skagway, Alaska, Final Report

Alaska Department of Health and Social Services, Division of Public Health (2001) Public health evaluation of exposure of Kivalina and Noatak residents to heavy metals from Red Dog Mine, Alaska

Bellinger DC (2004) Lead. Pediatrics 113(4):1016–1022

Bluemink E (2008) Blood lead levels in Alaska raise concern. Anchorage Daily News, Anchorage, AK, 3 March

Brumbaugh WG, May TW (2008) Elements in mud and snow in the vicinity of the DeLong Mountain Regional Transportations System road, Red Dog Mine and Cape Krusenstern National Monument, Alaska, 2005–2006. United States Geological Survey Scientific Investigations Report 2008-5040.

Canfield RL, Henderson CR, Cory-Slechta DA, Cox C, Jusko TA, Lanphear BP (2003) Intellectual impairment in children with blood lead concentrations below 10 micrograms per deciliter. N Engl J Med 348:1517–1526

Chiaradia M, Gulson BL, MacDonald K (1997) Contamination of houses by workers occupationally exposed in a lead–zinc–copper mine and impact of blood lead concentrations in the families. Occup Environ Med 54(2):117–124

Compañia Minera Antamina SA (2006) Antamina environmental monitoring plan. Document Reference No. DC.1.005

Code of Federal Regulations. Title 29 Part 1910.1027 (2007) Occupational safety and health standards: cadmium

Dames & Moore, Inc. Consulting (1983) Cominco Alaska Inc., Environmental baseline studies, Red Dog Project

Exponent Engineering and Scientific Consulting (2007) DeLong Mountain transport system fugitive dust risk assessment, vol I – Report. Prepared for Teck Cominco Alaska, Alaska. Document No. 8601997.007 5400 1107 SS15

Exponent Engineering and Scientific Consulting (2008) Draft fugitive dust risk management plan: Red Dog Operations, Alaska. Prepared for: Teck Cominco Alaska, Inc. Document No. 8601997.008 5800 0708 SS25

Ford J, Hasselbach L (2001) Heavy metals in mosses and soils on six transects along the Red Dog Mine haul road, Alaska. National Park Service Technical Report NPS/AR/NRTR-2001/38

Foulkes EC (1998) Biological membranes in toxicology. Taylor and Francis, Philadelphia, PA

Hasselbach L, Ver Hoef JM, Ford J, Neitlich P, Crecelius E, Berryman S, Wolk B, Bohle T (2005) Spatial patterns of cadmium and lead deposition on and adjacent to National Park Service lands in the vicinity of Red Dog Mine, Alaska. Sci Total Environ 348:211–230

Henson MC, Chedrese PJ (2004) Endocrine disruption by cadmium, a common environmental toxicant with paradoxical effects on reproduction. Exp Biol Med 229:383–392

Hu H, Aro A, Payton M, Korrick S, Sparrow D, Weiss ST, Rotnitzky A (1996) The relationship of bone and blood lead to hypertension – the normative aging study. JAMA 275(15):1171–1176

Kelley KD, Hudson T (2007) Natural versus anthropogenic dispersion of metals to the environment in the Wulik River area, western Brooks Range, northern Alaska. Geochem: Explor Environ Anal 7(1):87–96

Klaassen CD, Amdur MO, Doull J (1996) Casarett and Doull's toxicology: the basic science of poisons, 5th edn. McGraw-Hill, New York, NY

Kosnett MJ, Wedeen RP, Rothenberg SJ, Hipkins KL, Materna BL, Schwartz BS, Hu H, Woolf A (2007) Recommendations for medical management of adult lead exposure. Environ Health Perspect 115(3):463–471

Kral S (1992) Red Dog: Cominco's arctic experience pays off again. Mining Eng 44(1):43–49

Liles P (2006) Red Dog churns out record profits. Can Min J 127(4):26–29

Maret W, Sandstead HH (2006) Zinc requirements and the risks and benefits of zinc supplementation. J Trace Elem Med Biol 20:3–18

Murgueytio AM, Evans RG, Sterling DA, Clardy SA, Shadel BN, Clements BW (1998) Relationship between lead mining and blood lead levels in children. Arch Environ Health 53(6):414–423

O'Hara T, George JC, Blake J, Burek K, Carroll G, Dau J, Bennett L, McCoy CP, Gerard P, Woshner V (2003) Investigation of heavy metals in a large mortality event in caribou of Northern Alaska. Arctic 56(2):125–135

Pirkle JL, Kaufmann RB, Brody DJ, Hickman T, Gunter EW, Paschal DC (1998) Exposure of the U.S. population to lead, 1991–1994. Environ Health Perspect 106(11):745–750

Reed WR, Organiscak JA (2007) Haul road dust control. Coal Age 112(10):34–37

Roscoe RJ, Ball W, Curran JJ, DeLaurier C, Falken MC, Fitchett R, Fleissner ML, Fletcher AE, Garman SJ, Gergely RM, Gerwel BT, Gostin JE, Keyvan-Larijani E, Leiker RD, Lofgren JP, Nelson DR, Payne SF, Rabin RA, Salzman DL, Schaller KE, Sims AS, Smith JD, Socie EM, Stoeckel M, Stone RR, Whittaker SG, (2002) Adult blood lead epidemiology and surveillance: United States, 1998–2001. Morb Mortal Wkly Rep 51(SS11):1–10

Satarug S, Haswell-Elkins MR, Moore MR (2000) Safe levels of cadmium intake to prevent renal toxicity in human subjects. Br J Nutr 84:791–802

Tetra Tech (2008) Red Dog Mine extension: Aqqaluk project. Draft supplemental environmental impact statement. Prepared for: Teck Cominco Alaska, Alaska. http://www.reddogseis.com/

Thornton I, Davies DJ, Watt JM, Quinn MJ (1990) Lead exposure in young children from dust and soil in the United Kingdom. Environ Health Perspect 89:55–60

USEPA (1984) Draft environmental impact statement, Red Dog Mine project, northwest Alaska, vol I and II. U.S. EPA Document No. 910/9-84-122a

USEPA (1998) National air quality and emissions trends report, 1997. Office of Air and Radiation, U.S. EPA, Research Triangle Park

USEPA (2004) Air quality criteria for particulate matter final report. U.S. EPA Document No. 600/P-99/002aF-bF

USEPA (2007) Scoping document for the Red Dog Mine extension: Aqqaluk project supplemental environmental impact statement. U.S. Environmental Protection Agency Region 10, Seattle, WA

Wenzel S (2008) Adult blood lead epidemiology and surveillance: occupational exposures: Alaska, 1995–2006. State of Alaska Epidemiology Bulletin 2: January 23, 2008.

A Profile of Ring-hydroxylating Oxygenases that Degrade Aromatic Pollutants

Ri-He Peng, Ai-Sheng Xiong, Yong Xue, Xiao-Yan Fu, Feng Gao, Wei Zhao, Yong-Sheng Tian, and Quan-Hong Yao

Contents

1 Introduction 65
2 Classification of Ring-hydroxylating Oxygenases 66
3 Structural Investigations of Ring-hydroxylating Oxygenases 72
   3.1 The Structure of the β Subunit 73
   3.2 The Structure of the α Subunit 75
4 Ring-hydroxylating Oxygenases: Electron Transfer and Substrate Oxidation 78
5 Regioselectivity and Stereoselectivity of Ring-hydroxylating Oxygenases 81
6 Techniques for Improving Ring-hydroxylating Oxygenase Degradation Capabilities . 85
7 Summary 87
References 88

1 Introduction

Aromatic compounds are widely distributed in nature and range in size from low molecular mass compounds, such as phenols, to polymers such as lignin (Vaillancourt et al. 2006). As a result of the delocalization of their resonance structure, aromatic compounds are exceptionally stable (McMurry 2004). Because of the metabolic pathways they have evolved, microorganisms have an exceptional ability to utilize aromatic compounds as their sole source of energy and carbon (Pieper and Reineke 2001; Reineke and Knackmuss 1988).

Several aromatic compounds, such as polycyclic aromatic hydrocarbons (PAHs), are widespread in various ecosystems and are regarded as hazardous pollutants of great concern primarily because of their toxicity, including a tendency to induce

---

Q.-H. Yao (✉)
Shanghai Key Laboratory of Agricultural Genetics and Breeding, Agro-Biotechnology Research Institute, Shanghai Academy of Agricultural Sciences, 2901 Beidi Rd, Shanghai, People's Republic of China
e-mail: yaoquanhong_sh@yahoo.com.cn

mutagenicity and carcinogenicity. The use of microbial metabolic potential for eliminating aromatic pollutants such as PAHs provides a safe and economic alternative to their disposal in waste dump sites and to commonly used physico-chemical strategies. The ring-hydroxylating oxygenases (RHOs) play a key role in microbial biodegradation because they catalyze the first step in the degradation process. For example, naphthalene dioxygenase is particularly useful for dihydroxylating the low molecular weight PAHs (Peng et al. 2008). To date, more than 100 RHOs have been identified in different microorganisms; most of these RHOs have a broad substrate range and can catalyze diverse oxidative reactions. It is our goal in this review to offer a profile of the nature of these RHOs. Such information may be helpful to researchers who study or rely on RHOs for degrading a variety of pollutants, either naturally or in directed programs that threaten the modern environment.

RHOs are multi-component enzymes, comprising two or three protein components, and structurally consist of an electron transport chain (ETC) and an oxygenase (Ashikawa et al. 2006; Ferraro et al. 2005). The oxygenase components are either homooligomers (α_n) or heterooligomers ($\alpha_n\beta_n$) and in each case, the α subunit, called the large subunit, has two conserved regions, a Rieske [2Fe–2S] center and a non-heme mononuclear iron (Butler and Mason 1997; Mason and Cammack 1992). These oxygenase systems usually catalyze the oxidation of various aromatic compounds by introducing two hydroxyl groups, either in the *ortho*-position or in the *para*-position, by which a soluble electron transport chain is formed to harness the reductive power of NAD(P)H and activate molecular oxygen (Gibson 1971; Resnick et al. 1996). The RHO system catalyzes the oxidation of an arene bond to yield an arene *cis*-dihydrodiol (Jeffrey et al. 1975; Ziffer et al. 1973) and is thereby different from the known P450 monooxygenase systems that form *trans*-dihydrodiols (Raag and Poulos 1989).

In recent years, a tremendous accumulation of new sequence data has been developed for RHOs (Kweon et al. 2008). Identification and characterization of these RHOs has allowed more detailed studies of the enzyme systems to be undertaken. In this review of RHOs, we address the following key topics: (a) the relationships between different RHOs, (b) structural information on RHOs and their relationship to substrate specificity, (c) the catalytic mechanisms by which complex enzymatic oxidation reactions proceed, and (d) the techniques that enhance the pollutant degradation capabilities of RHOs.

2 Classification of Ring-hydroxylating Oxygenases

RHOs of bacterial origin can catalyze the oxidation of a variety of hydrophobic, mainly aromatic substances by the insertion of one or two hydroxyl groups. They are multi-component enzymes containing two or three protein components constituting an ETC and an oxygenase. The terminal oxygenase is known to be the catalytic component involved in the transfer of electrons to oxygen molecules. The ETC that transfers reducing equivalents from NAD(P)H to the oxygenase components consists of either a flavoprotein reductase or a flavoprotein reductase and a ferredoxin. Although both oxygenase and ferredoxin belong to Rieske-type proteins, the phylogenetic tree that describes the relationships between oxygenases is more complicated than that of ferredoxins, because oxygenases exhibit a greater diversity in their substrate specificity and quaternary structure (Schmidt and Shaw 2001).

Since the discovery of ferredoxins in the early 1960s, the number of identified proteins that contain [Fe–S] clusters has been greatly expanded; these are referred to as [Fe–S] proteins (Fontecave and Ollagnier-de-Choudens 2008; Rieske et al. 1964). Such Rieske proteins contain so-called Rieske [2Fe–2S] clusters and play important roles in many biological electron transfer reactions (Cosper et al. 2002; Mason and Cammack 1992; Schäfer et al. 1996).

Sequence analyses of various Rieske proteins revealed that, in general, they all possess a common homologous region having the sequence: -C-X-H-X15–17-C-X-X-H- (Castresana et al. 1995; Carrell et al. 1997; Mason and Cammack 1992; Neidle et al. 1991). In the Rieske cluster, there is an asymmetric iron–sulfur core with the S^γ atom of each of the two cysteine residues coordinated to one iron site, and the N^δ atom of each of the two histidine residues coordinated to the other iron site. This asymmetric ligation results in some unique redox and spectroscopic properties. Structural variation in the vicinity of the clusters, such as hydrogen bond networks (Denke et al. 1998; Guergova-Kuras et al. 2000; Schröter et al. 1998), the iron–histidine bond length (Cosper et al. 2002), and polypeptide dipoles (Colbert et al. 2000), may be correlated with reduction potential. Although the RHOs exhibit quite different catalytic functions, their possession of a common sequence motif suggests that they are in some way related, presumably by divergent evolution from a common ancestor (Asturias et al. 1995; Nakatsu et al. 1995; Schmidt and Shaw 2001).

The RHOs have initially been classified into three classes by Batie et al. (1991), based on the number of constituent components and the nature of the RHO redox centers. Class I RHOs are composed of two protein components: a reductase and an oxygenase. This class contains two subtypes. In type IA, the oxygenase possesses a homomultimeric quaternary structure, whereas type IB possesses a heteromultimeric quaternary structure composed of α and β subunits. The α subunits bear the catalytic center of these oxygenases. In addition, the reductases in the type IA contain a flavin mononucleotide (FMN)-binding domain, whereas in the type IB they contain a flavin adenine dinucleotide (FAD)-binding domain. Some RHOs, such as phthalate dioxygenase from *Burkholderia cepacia* (AF095748) (Chang and Zylstra 1998); 3-chlorobenzoate-3,4-dioxygenase from *Alcaligenes* sp. (Q44256) (Nakatsu et al. 1995); and toluene sulfonate methylmonooxygenase from *Comamonas testosteroni* (U32622) (Junker et al. 1997) belong to type IA, whereas some benzoate-1,2-dioxygenases, such as benzoate-1,2-dioxygenase from *Acinetobacter* sp. ADP1 (AF009224) (Neidle et al. 1991), and benzoate-1,2-dioxygenase from *Pseudomonas putida* (TOL plasmid, M64747) (Harayama et al. 1991) belong to type IB.

Class II and class III RHOs contained three protein components: a flavoprotein reductase, a ferredoxin, and a terminal oxygenase. Reducing equivalents are transferred from NAD(P)H to the terminal oxygenase through a flavoprotein reductase and a ferredoxin. Both classes are distinguished on the basis of the flavoprotein reductase. In class II RHOs, the flavoprotein reductase contains three domains: an FAD-binding domain, an NADH-binding domain, and a C-terminal domain, whereas the corresponding reductase in class III RHOs has an additional plant-type [2Fe–2S] cluster domain. Typical class III RHOs are naphthalene dioxygenase from *P. putida* (strain NCIB9816-4) (Kurkela et al. 1988) and naphthalene dioxygenase from *P. putida* (strain G7) (Simon et al. 1993). Based on the small protein ferredoxin structure, class II RHOs are further divided into two subtypes: type IIA and type IIB. In type IIA the ferredoxin contains a plant-type [2Fe–2S] cluster and in type IIB a Rieske-type [2Fe–2S] cluster. For example, dioxin dioxygenase from *Sphingomonas* sp. (Bünz and Cook 1993) belongs to type IIA but some benzene dioxygenase (Irie et al. 1987) and biphenyl dioxygenase (Erickson and Mondello 1992; Fukuda et al. 1994) enzymes belong to type IIB (Table 1).

The Batie classification system has been widely accepted because it was based on the composition of the electron transport chain and is capable of systematically describing the relationship of RHOs. However, in recent years, with the identification and characterization of more oxygenases, it is clear that the Batie classification system chain cannot credibly be used to classify all oxygenases. For example, the carbazole 1,9a-dioxygenase system from *Pseudomonas* sp. (strain CA10) can be grouped into class IA, based on the terminal oxygenase CarAa, because it is composed of a homomultimer, which is typically characteristic for class IA oxygenases in the Batie's system. However, based on the ferredoxin reductase CarAd, this dioxygenase system would be classified into class III, because it contains a plant-type [2Fe–2S] cluster and a flavin adenine dinucleotide-binding domain (Nam et al. 2002; Sato et al. 1997).

A second method of classification introduced by Nam et al. (2001) places RHOs into families based only on the sequence homology of the terminal oxygenase components. By comparing their sequences pairwise, the RHOs can be classified into four groups, described below.

Oxygenase components included in group I have only large α subunits, whereas other oxygenase components in groups II, III, and IV are composed of different subunits (large subunit α and small subunit β). Group I RHOs contain wide-ranging oxygenases that share low homology, while groups II, III, and IV RHOs contain typical benzoate/toluate dioxygenases, naphthalene/polycyclic aromatic hydrocarbon dioxygenases, and benzene/toluene/biphenyl dioxygenases, respectively. There are some interesting characteristics in the consensus sequences of the Rieske-type [2Fe–2S] cluster-binding site and the Fe^{2+}-binding site in the terminal oxygenase. First, group I oxygenases have 16 or 18 amino acids between the first His and the second Cys in the Rieske-type [2Fe–2S] cluster that binds site sequences C-X-H-X15–17-C-X-X-H, whereas groups II, III, and IV have 17 such amino acids. Second, the two His residues of the Fe^{2+}-binding site are separated by three or four amino acids in group I oxygenases, whereas they are separated by four in groups II and III and by four or five in group IV. In addition, two conserved sequence blocks (Gly-Asn at the Rieske cluster-binding site and Asn-Trp-Lys at the Fe^{2+}-binding site) give

Table 1 Batie classification scheme for ring-hydroxylating oxygenases (Batie et al.1991)

Class	Reductase	Ferredoxin	Oxygenase	Enzyme system
IA	FMN[2Fe–2S]$_P$		[2Fe–2S]$_R$ Fe$^+$	Phthalate dioxygenase (*Burkholderia cepacia*) Phenoxybenzoate dioxygenase (*Pseudomonas pseudoalcaligenes*) 3-Chlorobenzoate 3,4-Dioxygenase (*Alcaligenes* sp.)
IB	FAD[2Fe–2S]$_R$		[2Fe–2S]$_R$ Fe$^+$	Benzoate 1,2-dioxygenase (*Acinetobacter* sp.) 2-Oxo-1,2-dihydroquinoline 8-monooxygenase (*Pseudomonas putida*) Toluate 1,2-dioxygenase (*Pseudomonas putida*) Anthranilate dioxygenase (*Acinetobacter* sp.)
IIA	FAD	[2Fe–2S]$_P$	[2Fe–2S]$_R$ Fe$^+$	Dibenzofuran dioxygenase (*Sphingomonas* sp.) Pyrazon dioxygenase (*Pseudomonas* sp.)
IIB	FAD	[2Fe–2S]$_R$	[2Fe–2S]$_R$ Fe$^+$	Toluene dioxygenase (*Pseudomonas putida*) Benzene 1,2-dioxygenase (*Pseudomonas putida*) Biphenyl dioxygenase (*Pseudomonas* sp.) Cumene dioxygenase (*Pseudomonas fluorescens*)
III	FAD[2Fe–2S]$_R$	[2Fe–2S]$_R$	[2Fe–2S]$_R$ Fe$^+$	Carbazole 1,9a-dioxygenase (*Pseudomonas* sp.) Naphthalene dioxygenase (*Pseudomonas putida*) 2-Nitrotoluene dioxygenase (*Pseudomonas* sp.) Nitrobenzene dioxygenase (*Comamonas* sp.)

FMN: flavin mononucleotide; FAD: flavin adenine dinucleotide

a good clue for the classification of the RHOs. It was speculated that both blocks affect the interaction of α and β subunits, because they disappear in the oxygenases of group I that contain only an α subunit (Table 2).

Compared to the Batie classification system, the Nam classification system is simple and powerful. Dioxygenases can be easily classified after comparing them against the two standards: toluene dioxygenases (TodC1) from *P. putida* F1 (Subramanian et al. 1979) or benzene dioxygenase (BedC1) from *P. putida* ML2 (Irie et al. 1987). For example, the enzyme 2-oxo-1,2-dihydroquinoline

Table 2 Nam classification scheme for ring-hydroxylating oxygenases (Nam et al. 2001)

Group	Subunit	[2Fe–2S]-binding site	Fe^{2+}-binding site	Enzyme system
I	α	C-X-H-X(15–17)-C-X-X-H	D-X(2)-H-X(3-4)-H	2-Oxo-1,2-dihydroquinoline 8-monooxygenase Carbazole 1,9a-dioxygenase
II	α+β	C-X-H-X(10)-GN-X(5)-C-X-X-H	NWR-X(7)-D-X(2)-H-X(4)-H	Benzoate/toluate dioxygenases
III	α+β	C-X-H-X(10)-GN-X(5)-C-X-X-H	NWR-X(8)-D-X(2)-H-X(4)-H	Naphthalene/polycyclic aromatic hydrocarbon dioxygenases
IV	α+β	C-X-H-X(10)-GN-X(5)-C-X-X-H	NWR-X(8)-D-X(2)-H-X(4-5)-H	Benzene/toluene/biphenyl dioxygenase

8-monooxygenase (Y12655.1) from *P. putida* is difficult to classify using the Batie classification scheme, because, although it functions with an electron transport chain characteristic of the type-IB oxygenases, the sequence of this terminal oxygenase component is closely related to those of the type-IA oxygenases (Rosche et al. 1995, 1997). However, this enzyme can be easily classified into group I of the Nam classification scheme as a result of the low homology with TodC1 or BedC1. 7-Oxodehydroabietic acid dioxygenase, which is involved in diterpenoid degradation in *Pseudomonas abietaniphila* BKME-9 (DitA1), is also difficult to classify in the Batie classification scheme. This is because this enzyme contains a [3Fe–4S]-type ferredoxin, which is unusual for the oxygenases (Martin and Mohn 1999). But the DitA oxygenase was successfully grouped into group III in the Nam classification scheme because the terminal oxygenase had a sequence homology score of about 26 with TodC1 or BedC1 (Nam et al. 2001).

Recently, Kweon et al. (2008) introduced a new classification scheme for RHOs based on analyzing the terminal oxygenase and the ETC components as a whole. To effect enzyme classification by this approach, first, pairwise and multiple alignments of the terminal oxygenases were carried out with the default parameters (protein weight matrix, for example). The pairwise distance (PD) matrices were used to maximize the accuracy of classification. When the mean PD value approached 0.61, the maximum accuracy of classification is obtained and not much affected by the number of observed oxygenases. This indicated that a PD value of 0.61 is suitable as a criterion for grouping oxygenases. Moreover, phylogenetic data on the oxygenases being classified were integrated with the classification keys obtained from ETC components.

Oxygenase components of RHOs were phylogenetically classified into five distinct types, using the Kweon classification scheme, and all the members within each type share the same classification keys as regards ETC components. Type I represents a two-component RHO system that consists of an oxygenase and a C-type ferredoxin–$NADP^+$ reductase (FNR), with a C terminus of NAD domains.

Type II contains another two-component RHO system that consists of an oxygenase and an N-type FNR reductase having an N terminus of a flavin-binding domain. Type III represents a group of three-component RHO system that consists of an oxygenase, a [2Fe–2S]-type ferredoxin, and an N-type FNR reductase. Type IV represents another three-component system containing oxygenase, [2Fe–2S]-type ferredoxin, and glutathione reductase (GR)-type reductase. Type V stands for another different three-component system that consists of an oxygenase, a [3Fe–4S]-type ferredoxin, and a GR-type reductase (Kweon et al. 2008) (Table 3). Genes coding for ETC components are not always closely positioned with oxygenase genes (genetic discreteness) and limited numbers of ferredoxin and reductase components are shared by multiple oxygenases (numerical imbalance). Therefore, it was a challenge to find a cognate ETC partner that can function in cooperation with these "incomplete" oxygenase components (Armengaud and Timmis 1997; Jones et al. 2003; Romine et al. 1999; Stingley et al. 2004). Using estimates of PD value, Kweon devised a model to determine a suitable set of oxygenases designed to minimize the classification error, even if the corresponding ETC information was lacking. For example, the PD value can be used to differentiate DitA (*P. abietaniphila* BKME-9) (Martin and Mohn 1999) and NidA (*Rhodococcus* sp. strain I24) (Treadway et al. 1999) from the NahAc (*Pseudomonas* sp. NCIB9816-4) (Gibson et al. 1995) and PahAc (*P. putida* OUS82) (Takizawa et al. 1994).

Table 3 Kweon classification scheme for ring-hydroxylating oxygenases (Kweon et al. 2008)

Class		Subunit	Reductase	Ferredoxin	Enzyme system
I	Iα	α	FNR_C-type $FMN[2Fe-2S]_P$	None	Aniline dioxygenase (*Acinetobacter* sp. YAA) Aniline oxygenase (*Pseudomonas putida* UCC22)
	Iαβ	α+β	FNR_C-type $FAD[2Fe-2S]_R$	None	Phenoxybenzoate dioxygenase (*Alcaligenes* sp. BR60) Phenoxybenzoate dioxygenase (*Pseudomonas pseudoalcaligenes* POB310) Phthalate dioxygenase (*Burkholderia cepacia* DBO1) Toluene sulfonate monooxygenase (*Comamonas testosteroni* T-2)
II		α+β	FNR_N-type $FAD[2Fe-2S]_R$	None	2-Halobenzoate 1,2-dioxygenase (*Pseudomonas cepacia* 2CBS) Benzoate 1,2-dioxygenase (*Acinetobacter* sp. ADP1) Anthranilate dioxygenase (*Acinetobacter* sp. ADP1)
III	IIIα	α	FNR_N-type $FAD[2Fe-2S]_R$	[2Fe–2S]-type	Carbazole 1,9a-dioxygenase (*Pseudomonas resinovorans* CA10)

Table 3 (continued)

Class	Subunit	Reductase	Ferredoxin	Enzyme system
IIIαβ	α+β	FNR$_N$-type FAD[2Fe–2S]$_R$	[2Fe–2S]-type	Naphthalene dioxygenase (*Pseudomonas* sp. 9816-4)
				3,4-Dihydroxyphenanthrene dioxygenase (*Alcaligenes faecalis* AFK2)
				PAH dioxygenase (*Pseudomonas putida* OUS82)
				Naphthalene dioxygenase (*Ralstonia* sp. U2)
				Salicylate 5-hydroxylase (*Ralstonia* sp. U2)
IV	α+β	GR-type FAD	[2Fe–2S]-type	Carbazole dioxygenase (*Sphingomonas* sp. CB3)
				Dioxin dioxygenase (*Sphingomonas* sp. RW1)
				Biphenyl dioxygenase (*Rhodococcus* sp. RHA1)
				Toluene dioxygenase (*Pseudomonas putida* F1)
				Biphenyl 2,3-dioxygenase (*Pseudomonas* sp. LB400)
				Biphenyl dioxygenase (*Pseudomonas pseudoalcaligenes* KF707)
V	α+β	GR-type FAD	[3Fe–4S]-type	Phenanthrene dioxygenase (*Nocardioides* sp. KP7)
				Phthalate dioxygenase (*Terrabacter* sp. DBF63)
				Phthalate dioxygenase (*Mycobacterium vanbaalenii* PYR-1)

FNR: ferredoxin–NADP$^+$ reductase; GR: Glutathione reductase; PAH: Polyaromatic hydrocarbons

The Kweon classification system has responded dynamically to the growing pool of RHO enzymes. As standard RHO samples increase, the classification system increases in objectivity and stability. In turn, this objectivity and stability of the classification system may help extend classification coverage to many other RHO enzymes.

3 Structural Investigations of Ring-hydroxylating Oxygenases

Most RHOs catalyze insertion of molecular oxygen into aromatic benzene rings to form arene oxides. This reaction requires an electron to be transported from an NAD(P)H reductase to the terminal oxygenase component. In the terminal oxygenase, the electron travels from the Rieske cluster to the mononuclear iron for use in catalysis. The active site pocket of the terminal oxygenase has been well described with respect to oxygen and substrate binding. A large amount of structural information relevant to RHOs is now available in the protein data bank (PDB). For example, the PDB retains information on the following RHOs: naphthalene dioxygenases from *Pseudomonas* sp. strain NCIB9816-4 (NDO-O_{9816-4}) (Kauppi et al. 1998), *Sphingomonas* strain CHY-1 (PHNI-O_{CHY-1}) (Jakoncic et al. 2007a), and *Rhodococcus* sp. strain NCIMB12038 (NDO-O_{12038}) (Gakhar et al. 2005), nitrobenzene dioxygenase from *Comamonas* sp. strain JS765 (NBDO-O_{JS765}) (Friemann et al. 2005), toluene 2,3-dioxygenase from *P. putida* strain F1 (TDO-O_{F1}) (Friemann et al. 2009), biphenyl dioxygenase from *Rhodococcus* sp. strain RHA1 (BPDO-O_{RHA1}) (Furusawa et al. 2004), cumene dioxygenase from *Pseudomonas fluorescens* strain IP01 (CDO-O_{IP01}) (Dong et al. 2005), 2-oxoquinoline 8-monooxygenase from *P. putida* strain 86 (OMO-O_{86}) (Martins et al. 2005), and carbazole 1,9a-dioxygenase from *Pseudomonas resinovorans* strain CA10 (CARDO-O_{CA10}) (Nojiri et al. 2005). Structures of enzyme–substrate complexes and their relationship to known product regioselectivities suggest that binding orientation of the substrate in the active site is the primary determinant of product regiospecificity. It is also possible that the type of reaction catalyzed (e.g., monohydroxylation vs. dihydroxylation) may be derived from the orientation of substrate binding at the active site.

Using gel-filtration analysis, the terminal dioxygenase of RHOs has been characterized as having an $\alpha_3\beta_3$ or α_3 configuration. Most enzymes exhibited an $\alpha_3\beta_3$ quaternary structure (Fig. 1a and b). The three α and β subunits form tight trimers, which bury a large part of their accessible surface area. The overall shape of the full hexameric complex resembles that of a mushroom, in which the stem consists of the β_3 subunits and the cap the α_3 subunits (Fig. 1a). The α subunit contains a substrate binding and catalytic domain having the mononuclear Fe(II) center and a Rieske domain with the [2Fe–2S] cluster. As the β subunit has no direct interaction with the active site, its main role is believed to provide only structural stability to the enzyme. Indeed, many dioxygenases, such as those of the phthalate family, lack the β subunit altogether.

3.1 The Structure of the β Subunit

Recent crystal structure studies show that nearly all oxygenases of the RHO type exist as $\alpha_3\beta_3$ multimers that have subunits arranged head-to-tail in α- and β-stacked planar rings. Only the oxygenases in group I of Nam classification scheme have identical α subunits. Detailed investigation of the sequence alignment has shown that group I oxygenases lack the Gly-Asn and Asn-Trp-Lys blocks, which are conserved in a Rieske-type [2Fe–2S] cluster-binding site and Fe^{2+}-binding site, respectively, in other group dioxygenases (Nam et al. 2001). There are some apparent structure differences between the monomeric and the heterohexameric oxygenases, such as subunit size and the size of the centrally located hole. Based on the crystal structure of CARDO-O_{CA10}, the doughnut shape is probably the common and typical one for terminal oxygenase components in the group I RHOs that have the α_3 configuration (Nojiri et al. 2005).

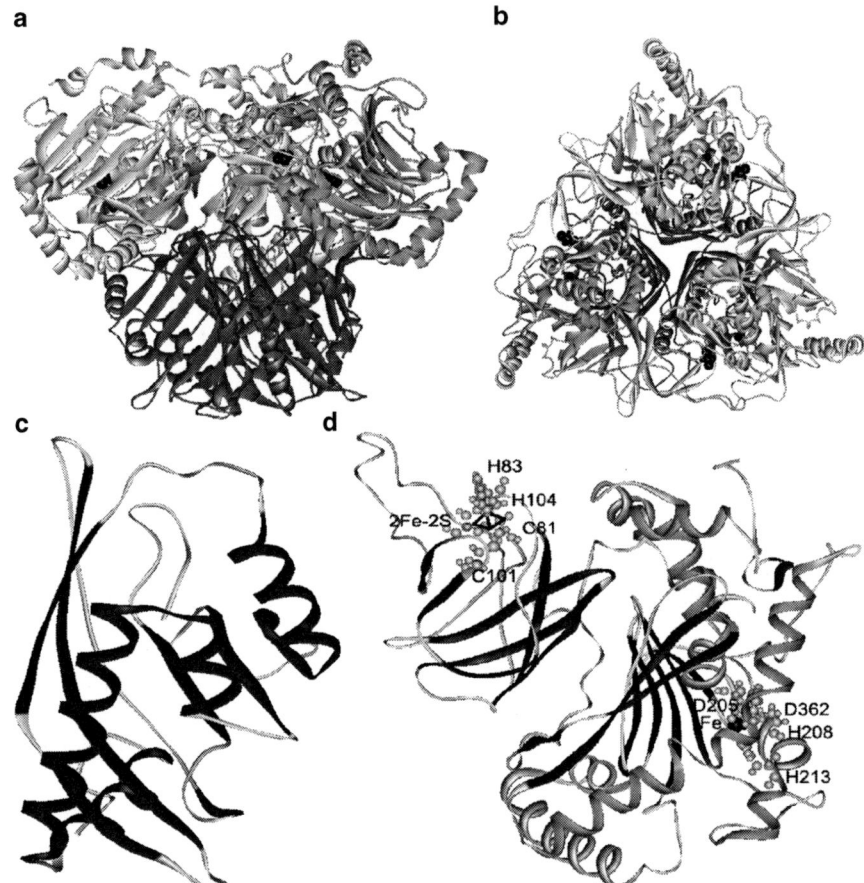

Fig. 1 Typical structure of ring-hydroxylating oxygenases. (**a**) Structure of the NDOα3β3 hexamer. The hexameric complex has a mushroom shape with the α subunits (*silver*) as the cap and the β subunits (*dim gray*) as the stem. The Rieske center and the catalytic iron are colored black. (**b**) This view is taken along the molecular threefold axis of the NDOα3β3 hexamer. (**c**) The tertiary structure of the β subunit of NDO-O. It is dominated by a long twisted six-stranded mixed β-sheet wrapped around three α helices. (**d**) The secondary structure of the α subunit of NDO-O. The α subunit is composed of a Rieske domain with the [2Fe–2S] cluster (*black*) and a catalytic domain with the mononuclear iron (*black*). The Rieske domain is dominated by three separate anti-parallel β-sheet structures and arranged in a sandwich topology. In the Rieske [2Fe–2S] center, one Fe atom is coordinated by the S^γ atoms of Cys81 and Cys101, whereas another is coordinated by the $N^{\delta 1}$ atoms of His83 and His104. The catalytic domain is dominated by a seven- to nine-stranded antiparallel β-sheet and the iron center was coordinated by the ligands His208, His213, and Asp362

In spite of low amino acid sequence identity, the β subunit shares the same global structural pattern, i.e., a funnel-shaped conical cavity. The tertiary structure is dominated by a long twisted six-stranded mixed β-sheet wrapped around three α helices. The central part of the contact area with other β subunits in the trimer is formed as a pleated sheet, whereas the α helices are located mainly on the outer part of the stem (Fig. 1c). The last four residues in the C-terminal coil are deeply anchored inside the core of the conical-shaped funnel by a hydrogen bonding network having strictly conserved arginine residues (Jakoncic et al. 2007a). The most significant structural difference of β subunits is detected in a long extended loop region and the N-terminal region. In NDO-O_{9816-4}, the long loop formed by residues Ser68-Met85 is involved in β subunit interactions and has interactions with the Rieske domain of the α subunit, as well. Residue Ser75 forms a main chain hydrogen bond to the main chain of residue Glu92 of the Rieske domain (Kauppi et al. 1998). In this region, the β subunit structure of NBDO-O_{JS765} is similar to NDO-O_{9816-4} (Friemann et al. 2005), but the PhnI secondary structure is closest to the structure of CDO-O_{IP01} (Dong et al. 2005) and BPDO-O_{RHA1} (Furusawa et al. 2004; Jakoncic et al. 2007a). The residues at the N terminus are also involved in trimer interactions between the β subunits.

Because they lack the β subunit, the structures of the regions for interaction between CARDO-O_{CA10} subunits are markedly different from that of NDO-O_{9816-4}. The NDO-O_{9816-4} interacts with the loop between β1 and β2 of the neighboring β subunit, whereas the CARDO-O_{CA10} interact with the α 11-3 of the neighboring α subunit by hydrophobic interaction with several H-bond networks via water molecules (Nojiri et al. 2005). In OMO-O86, the dimer is built up by the special small C-terminal domain (Martins et al. 2005). There is no similar structure in the α subunit of any described heterohexameric oxygenases.

3.2 The Structure of the α Subunit

The α subunit is composed of two domains: the Rieske domain with the [2Fe–2S] cluster and the catalytic domain with mononuclear iron.

The Rieske domain presents basically the same quaternary structure, which is dominated by three separate anti-parallel β-sheet structures, and is arranged in a sandwich topology. In NDO-O_{9816-4}, two hairpin structures protrude roughly perpendicular to the β-sheet and form two fingers that hold the [2Fe–2S] center at their tips. The first finger is formed by β7 and β8 and two iron ligands in the Rieske [2Fe–2S] center are positioned in the loop between the two strands. The second finger is composed of β10–β11–β12 and another pair of iron ligands is sited in the loop between β10 and β11. In the Rieske [2Fe–2S] center, one Fe atom is coordinated by the S^γ atoms of residue Cys81 and Cys101, while another is coordinated by the $N^{\delta 1}$ atoms of residues His83 and His104 (Kauppi et al. 1998) (Fig. 1d). All of these residues are absolutely conserved within RHOs. The two sulfide ions bridge the two iron ions and form a flat rhombic arrangement.

Although the structure of the Rieske domain in all RHOs has a similar fold, there are two noteworthy differences. The first difference is in the region containing ligands for the Rieske cluster. In these regions, each oxygenase subunit has a particular β strand; for example, β5 in the CARDO-O_{CA10}, β9 in the NDO-O_{9816-4}, and β7 in the BPDO-O_{RHA1}. The β5 strand of the CARDO-O_{CA10} constitutes one β-sheet clearly separated from the basal β-sandwich structure that is formed by the other two sheets. However, in the structures of NDO-O_{9816-4} and BPDO-O_{RHA1}, the corresponding regions form a loop, which extends to the basal β-sandwich structure. Since the long loop is involved in interactions with the respective β subunit of the oxygenase, the CARDO-O_{CA10} lacks the corresponding region (Nojiri et al. 2005). The second difference is the size and position of another long loop at the tip of the Rieske domain, which shields the [2Fe–2S] cluster from the solvent, and is positioned to interact with the catalytic domain from the adjacent α subunit. The loops are more extensive in NDO-O_{9816-4} and BPDO-O_{RHA1} than in CARDO-O_{CA10}. The difference in this region is thought to affect the interaction between the catalytic subunits, which may correlate with the difference between the α3 and α3β3 configurations. Both of these above-mentioned loop regions cannot be found in the Rieske domain of Rieske-type ferredoxins; for example, in BDO-F (Colbert et al. 2000) and CARDO-F (Nam et al. 2005), which exist as monomeric proteins.

There is an H-bond network between the Rieske [2Fe–2S] center and main chain in the oxygenases. Previous studies have indicated that the H-bond with the Rieske [2Fe–2S] center is important in determining the redox potential of the electron transfer proteins (Denke et al. 1998; Kolling et al. 2007). For example, in NDO-O_{9816-4} there are H-bonds from the main chain nitrogen of residues Arg84 and Lys86 to one of the sulfide ions and from the main chain nitrogen of residues His104 and Trp106 to the other sulfide ion. The four iron-ligating side chains are hydrogen bonded in a second coordination shell. Both histidine ligands are H-bonded to carboxylate residues in a neighboring subunit. Asp205 $O^{δ2}$ binds to the $N^{ε2}$ of His104 and Glu410^{δ2} binds to the $N^{ε2}$ of His83. The $S^γ$ atoms of the cysteinyl ligands Cys81 and Cys101 bind to the main chain nitrogen atoms of His83 and Tyr103, respectively (Kauppi et al. 1998). In higher redox potential Rieske cluster-containing proteins, such as mitochondrial cytochrome bc_1 (+300 mV), the side-chain atoms of the two cluster-proximal residues, Ser163 $O^γ$ and Tyr165 $O^η$, formed H-bonds with the cluster sulfide S1 and $S^γ$ of Cys139, respectively (Brugna et al. 1999; Schröter et al.1998; Trumpower 1981; Trumpower and Gennis 1994).

In most RHOs, the serine residue is replaced with tryptophan, but the tyrosine residue that forms an H-bond with the first cysteine is conserved, except for NDO-O_{9816-4}. Biochemical and biophysical studies of wild-type Rieske iron–sulfur protein from *Rhodobacter sphaeroides* and the Tyr156 → Phe mutant have demonstrated that mutation of the residues involved in hydrogen bonding with the cluster will result in a 45 mV decrease in the midpoint potential of the protein (Guergova-Kuras et al. 2000). However, the high-resolution X-ray crystal structures of the Tyr156 → Phe mutant protein reveals that the structure of the mutant is nearly identical to that wild-type protein. Hence, the measured 45-mV decrease in the midpoint potential in the Tyr156 → Phe mutant protein, relative to the wild type, is only the consequence of the loss of the hydrogen bond between Phe156 and the cluster ligand Cys129. Similarly, the measured 97-mV decrease in the midpoint potential in the Ser154 → Ala mutant protein can be attributed to be a direct consequence of the loss of the hydrogen bond between Ser154 and a bridging sulfur atom in the [2Fe–2S] cluster (Ugulava and Crofts 1998). The disulfide bond in the immediate vicinity of the [2Fe–2S] cluster is found in bc_1; the removal of this disulfide bond led to a protein with altered structure and a significant drop in the midpoint potential (Merbitz-Zahradnik et al. 2003). However, in the same region there is no disulfide bond among the known structures of the terminal oxygenases and ferredoxins of RHOs. The loss of disulfide bonding may result in a lower redox potential.

The catalytic domain is dominated by a seven- to nine-stranded anti-parallel β-sheet, which has tight connections between strands on the side bundling against the Rieske domain and has flexible insertions between the strands on the other side. The insertions contain ligands for mononuclear iron and form a catalytic pocket. A long broken helix covers one side of the sheet. In NDO-$O_{9816\text{-}4}$, the long helix α10 is broken into two parts at residues 353–355 that loop out from the helix (Kauppi et al. 1998). The catalytic pocket is formed by four structural motifs, three of which are α-helices, and the remaining motif is a β-sheet. A long cavity, extending from the surface to the anti-parallel β-sheet, provides substrates with access to the iron ion. Two flexible loops, LI and LII, are thought to act as lids covering the channel to the active site. Because the loops determine the pocket length, they may play a key role in the substrate selectivity of the enzyme (Jakoncic et al. 2007b) (Fig. 2). The sequence alignment of related enzymes shows that residues lining the catalytic pocket are well conserved among them. In NDO-$O_{9816\text{-}4}$, the mononuclear iron at the center of the catalytic pocket is coordinated by one carboxyl oxygen atom of Asp362, the $N^{\varepsilon 2}$ atoms of His208 and His213, and a water molecule. The geometry can be described as a distorted octahedral bipyramid with one ligand missing (Kauppi et al. 1998). This 2-His-1-carboxylate facial triad motif leaves the face of the iron exposed to the large hydrophobic active site, creating a catalytic platform wherein oxygen can bind and subsequently react with substrate to form product through a variety of reactions. Substitution of an alanine at position 362 in NDO-$O_{9816\text{-}4}$ completely eliminated enzyme activity (Parales et al. 2000). Site-directed mutagenesis of the corresponding histidines in toluene dioxygenase also resulted in inactive enzymes (Jiang et al. 1996).

The catalytic pocket can be divided into three regions, distal, central, and proximal, depending on the distance to the mononuclear iron atom. The central region consists mainly of hydrophobic residues (Jakoncic et al. 2007b). This hydrophobic environment around the substrate-binding pocket seems reasonable, since the enzyme prefers aromatic compounds as its primary substrates.

In NDO-$O_{9816\text{-}4}$, the Rieske [2Fe–2S] center is located 43.5 Å from the mononuclear iron center within a single α subunit, but is only 12 Å from the mononuclear iron of the catalytic domain in an adjacent α subunit, within the hexamer (Kauppi et al. 1998). It has been found that a large depression on the surface of the catalytic domain receives the Rieske domain from the adjacent α subunit, which places the [2Fe–2S] center in the right conformation with respect to the catalytic iron. Completely conserved among oxygenases is an aspartic acid buried in the large depression at the junction of the Rieske domain and the catalytic domain of the neighboring α subunit; this provides a bridge between these domains. In NDO-$O_{9816\text{-}4}$, Asp205 binds to His208, a ligand to the catalytic iron, and to His104, a ligand to the Rieske center in the adjacent α subunit (Kauppi et al. 1998). The

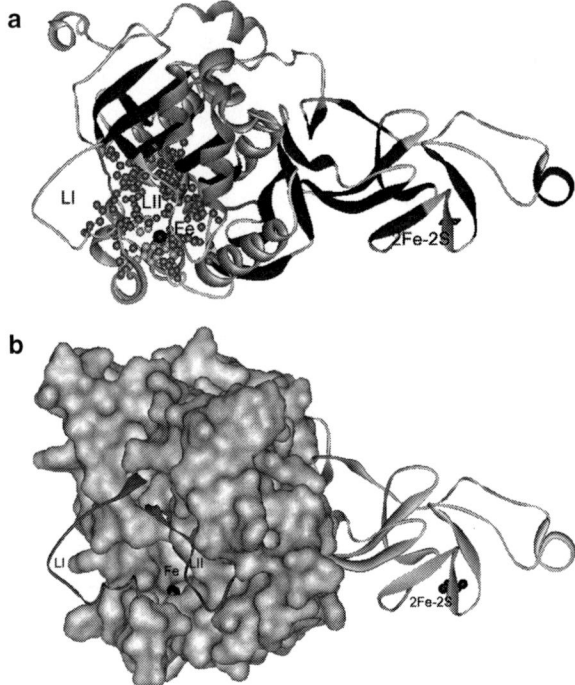

Fig. 2 The catalytic pocket in the PhnI α subunit: (**a**) The pocket is formed primarily by hydrophobic amino acids (*dim gray*). Two flexible loops, LI and LII, are thought to act as lids covering the channel to the active site. The iron was coordinated by the amino acid His207, His212, and Asp360 (*silver*), creating a catalytic platform where oxygen can bind and subsequently react with substrate. (**b**) Surface envelope of the catalytic pocket. From the surface, a narrow gorge is formed to provide substrates with access to the iron ion (*black*)

replacement of this aspartic acid by Ala, Glu, Gln, or Asn resulted in a totally inactive enzyme, suggesting that it is essential either directly in electron transfer or in positioning the two adjacent α subunits to allow effective electron transfer (Parales et al. 1999).

4 Ring-hydroxylating Oxygenases: Electron Transfer and Substrate Oxidation

Electron paramagnetic resonance (EPR) studies have shown that only limited amounts of mononuclear ferrous iron are oxidized at the active site when oxygen alone is allowed to react with the fully reduced NDO-$O_{9816\text{-}4}$. The role of substrate in the O_2 activation process was investigated by using benzoate 1,2-dioxygenase; results indicated that when substrate is present, the electron transfer reaction increased to a rate significantly faster than the enzyme turnover number (Wolfe et al. 2001). Frequently, substrate binding induces significant conformational changes around the active site of most RHOs and makes the binding site available. In NBDO-OJS765, substrate binding induces a change in the distance between the iron and oxygen atoms of the Asp360; this change is from 1.9 and 2.6 Å to 2.2 and 2.3–2.4 Å, and Asp360 coordinates the iron bidentately (Friemann et al. 2005). Similar bidentate conformations are present in structures of NDO-O_{9816-4}, with the Asp362 coordinating the iron monodentately only in the original structure. The crystal structures of reduced and oxidized NDO-O_{9816-4} reveal that dioxygen is bound side-on, close to the mononuclear iron at the active site. Because the dioxygen molecule is in a complex with the substrate and the dioxygen, it is positioned to attack the double bond of an aromatic substrate. Such a reaction explains the distinctive *cis*-stereospecific addition of both oxygen atoms to substrates by NDO-O_{9816-4}; these additions produce a *cis*-dihydrodiol (Karlsson et al. 2003).

The oxidation reaction catalyzed by RHOs is tightly regulated by the oxidation state of metal centers. For example, despite having high concentrations of the two substrates, naphthalene and O_2, the resting NDO-O_{9816-4} that retains a fully oxidized Rieske cluster and a partially reduced (Fe^{2+}) mononuclear center does not appear to catalyze substrate dioxygenation. However, upon reduction of the Rieske cluster and exposure to naphthalene and O_2, NDO-O_{9816-4} activity rapidly increases (Wolfe et al. 2001). The failure of the reduced mononuclear iron to react with O_2 in the resting NDO-O_{9816-4} accounts for the absence of an additional reducing equivalent. In OMO-O_{86}, the active site refuses the transmission of dioxygen to the mononuclear iron when it binds with the substrate. However, the situation changes when the active site senses the reduction of the Rieske center. Reduction of the Rieske center triggers the displacement of both the mononuclear iron and His221 away from the substrate (around 0.8 Å). Once protonated, His108 attracts Asp218, which enrolls His221 in a conformational change that alters the active site geometry and creates a pathway for dioxygen and a new coordination site at the mononuclear iron. The redox-coupled geometric rearrangement of mononuclear iron has been suggested to control end-on versus side-on binding of oxygen and possibly alteration of O_2 affinity (Martins et al. 2005).

There is a debate on the source of electrons and the timing of electron delivery during catalysis. The first possibility is that the Rieske cluster and the mononuclear iron together donate two electrons to O_2 to form an Fe(III) hydroperoxy intermediate, or form an Fe(V)–oxo–hydroxo species by O–O bond cleavage, when attacking the aromatic substrate (Wolfe et al. 2001) (Fig. 3). The second is that the reductase in the RHO system directly or indirectly donates another electron, before the attack on the substrate, to yield either an Fe(II)–(hydro)peroxo or an Fe(IV)–oxo–hydroxo species (Tarasev and Ballou 2005).

Recent studies on benzoate 1,2-dioxygenase have shown that an enzyme containing a fully oxidized Rieske cluster can utilize hydrogen peroxide to form a *cis*-diol product. The observation that the two-electron reduced oxygenase component can carry out a single turnover, when exposed to O_2, suggests that H_2O_2 may be able to provide both the electrons and the oxygen in a "peroxide shunt" reaction (Fig. 3).

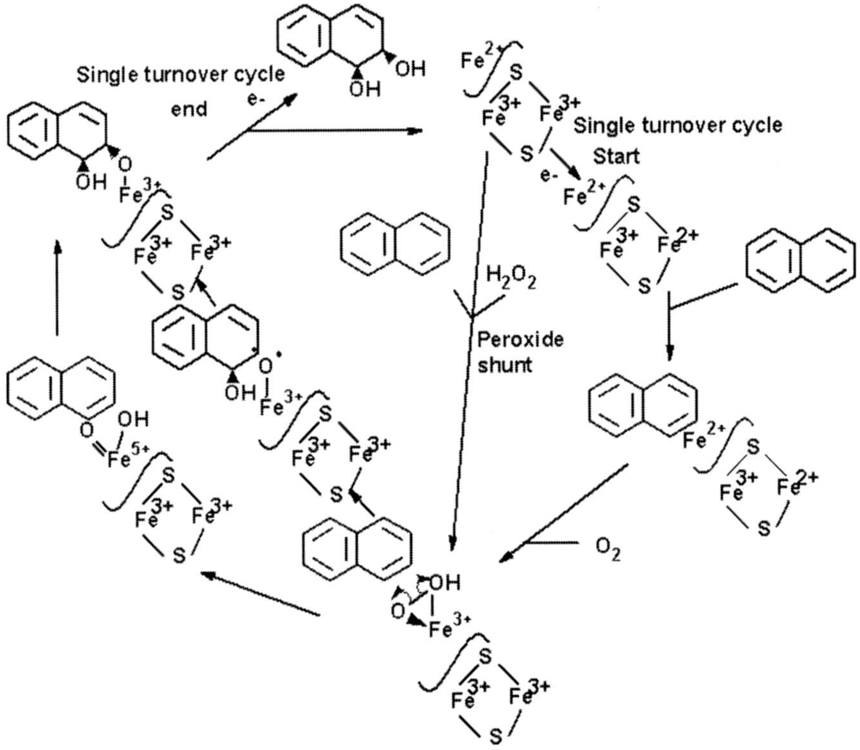

Fig. 3 Single-turnover and peroxide shunt reaction cycles of NDO. Possible radical-based mechanisms for *cis*-dihydroxylation that involves an Fe(III) hydroperoxy intermediate that then either forms an Fe(V)–oxo–hydroxo species by O–O bond cleavage or reacts directly with the aromatic substrate

However, at the end of the reaction, both metal centers were oxidized and the product was tightly bound to the active site. Thus, substrate (or product after a turnover cycle) must effectively limit the reaction to one turnover (Kovaleva et al. 2007; Neibergall et al. 2007).

Two research groups have unveiled the structure of the reductase and ferredoxin, and dioxygenase and ferredoxin complexes, opening new avenues for research into how the electrons are transferred from the reductase to the ferredoxin and from the ferredoxin to the oxygenase. Biochemical results showed that the interaction between ferredoxin (BphA3) and NADH-ferredoxin reductase (BphA4) was regulated in a redox-dependent manner. In the crystal of the BphA3–BphA4 complex, BphA3 binds to one subunit of homodimeric BphA4 at the FAD-binding and C-terminal domains. The crystal structures of these reaction intermediates prove that each elementary electron transfer induces a series of redox-dependent conformational changes in BphA3 and BphA4. The redox-dependent butterfly-like movement of the isoalloxazine ring of FAD, and rotation of the NAD-binding and C-terminal domains in BphA4, appears to be a prerequisite for the binding of BphA3. In the

BphA3–BphA4 complex, a side-chain rotation of His66 in BphA3 was induced to interact with Trp320 in BphA4. Thus, the electron is directly transferred from the flavin (gated by a Trp residue) to the Rieske cluster of the ferredoxin through His66. In addition, the conformational changes induced by the preceding electron transfer will induce the next electron transfer. The interaction of electron transfer and induced conformational changes seems to be critical to the sequential electron transfer reaction from NADH to ferredoxin (Senda et al. 2007).

By determining the crystal structure of the CARDO-O_{J3}–ferredoxin complex, Nojiri et al. (2005) provided an interpretation of intercomponent electron transfer between two Rieske [2Fe–2S] clusters of ferredoxin and oxygenase. Three molecules of CARDO-F bind to the subunit boundary of one CARDO-O trimeric molecule, and specific binding created by electrostatic and hydrophobic interactions with conformational changes suitably aligns the two Rieske clusters for electron transfer (Ashikawa et al. 2006).

5 Regioselectivity and Stereoselectivity of Ring-hydroxylating Oxygenases

The mononuclear iron of RHOs provides the platform for catalysis, although it alone cannot control substrate specificity and product regio- and stereoselectivity. Structural studies of known RHOs reveal that the orientation of substrate binding at the active site is the primary determinant of product regio- and stereo-selectivity. If multiple and equally favorable orientations of substrate are allowed at the active site, but one orientation places the more reactive atom closer to the mononuclear iron, there may be only one reaction product produced. The best result for each substrate was considered to be that with the lowest binding energy, after final docking. For example, in one of the lowest energy positions, indole is bound with its C2 and C3 atoms facing mononuclear iron, and N1 is positioned to make a favorable hydrogen bond with the carbonyl oxygen atom of Asp205 (of NDO-O_{9816-4}). It is the most suitable position for dihydroxylation by NDO. Similar studies with naphthalene and biphenyl indicated that the lateral positions that have low binding energies are the most suitable for yielding *cis*-dihydrodiol catalyzed by NDO (Carredano et al. 2000).

The active site amino acids that control substrate specificity and product regio- and stereo-selectivity were determined by site-directed mutagenesis methods. In NDO-O_{9816-4}, 17 residue side chains were identified to contribute to the overall topology of the substrate cavity, mainly by making a hydrophobic surface suitable for interactions with aromatic substrates (Carredano et al. 2000; Table 4). Among them, Phe 352 was found to be critical in determining the pocket shape and the regiospecificity of product. For example, when Phe352 was replaced with valine, the enzyme had the opposite regioselectivity with biphenyl and phenanthrene and a slight change in enantioselectivity with naphthalene and anthracene (Fig. 4a–c). In addition, the opposite enantiomers of biphenyl *cis*-3,4-dihydrodiol

Table 4 Structurally analogous residues that line the pocket of various ring-hydroxylating oxygenases

Position to catalytic Fe	Group III				Group V			Group I	
	NDO-O$_{9816-4}$	NBDO-O$_{JS765}$	PhnI	NDO-O$_{12038}$	BPDO-O$_{RHA1}$	TDO-O$_{F1}$	CDO-O$_{IP01}$	OMO-O$_{86}$	CARDO-O$_{CA10}$
P	Asn201	Asn199	Asn200	Asn209	Gln217	Gln215	Gln227	Asn215	Asn217
P	Phe202	Phe200	Phe201	Phe210	Phe218	Phe216	Phe228	Leu302	Leu270
								Gly216	Gly178
P	Val203	Val201	Val202	Val211	Cys219	Cys217	Cys229	Phe217	Phe179
C	Gly204	Gly202	Gly203	Gly212	Ser220	Ser218	Ser230	–	–
C	Ala206	Gly204	Gly205	Ala214	Met222	Met220	Met232	Asn219	Pro181
C	Val209	Val207	Val208	Thr217	Ala225	Ala223	Ala235	Ile222	Ile184
C	Leu217	Leu215	Leu216	Val225	Ile234	Ile232	Val244	Leu238	Leu200
D(LI)	Phe224	Leu222	Leu223	Phe293	Leu274	Leu272	Leu284	Val231	Ala199
D(LI)	Leu227	Leu225	Leu226	Phe236	Pro250	Pro248	Leu259	Pro239	Pro201
C	Gly251	Gly249	Gly251	Gly252	Gly266	Gly264	Gly276	–	Ile262
D(LII)	Leu253	Phe251	Ile253	Ile254	Tyr268	Tyr266	Phe278	Phe267	–
D(LII)	Val260	Asn258	Ile260	Met309	Ile278	Ile276	Ile288	Trp307	Phe275
C	His295	Phe293	His293	Phe293	Ala311	Val309	Ala321	Tyr292	Ala259
P	Asn297	Asn295	Asn295	His295	His313	His311	His323	Thr294	Leu270
C	Leu307	Leu305	Leu305	Phe307	Leu323	Leu321	Leu333	Gln314	Val272
C	Ser310	Ser308	Thr308	Phe320	Ile326	Ile324	Ile336	Trp307	Phe275
C	Phe352	Ile350	Phe350	Phe362	Phe368	Phe366	Phe378	Asn362	Asn330
C	Trp358	Trp356	Leu356	Phe368	Phe374	Phe372	Tyr384	Phe361	Phe329
C	Ala407	Ala405	Phe404	Asp418	Phe267	Phe265	Phe273	Leu266	–

Relative position to the catalytic Fe: D, distal; C, central; and P, proximal. (–) No structurally equivalent residues were observed
Groups were classified according the Nam classification scheme

and phenanthrene cis-1,2-dihydrodiol were formed, in contrast to the enantiomer formed by the wild-type NDO (Parales et al. 2000). Mutation of analogous residues in NBDO-O$_{JS765}$ (Ju and Parales 2006), BPDO from *Sphingomonas yanoikuyae* B8/36 (Parales et al. 2000), and 2-nitrotoluene 2,3-dioxygenase from *Acidovorax* sp. strain JS42 (Lee et al. 2005) also altered product regioselectivity. Replacement of other residues positioned near the active site iron has also allowed the enzyme to alter regio- and enantio-selectivities to some degree. For example, an NDO mutation (Ala206I) formed significantly more phenanthrene cis-1,2-dihydrodiol than did the wild type. In the presence of the F352I mutation, changes at positions 206 and 295 also affected enantioselectivity (Parales 2003).

Recent biochemical studies showed that the dioxygenase from *Sphingomonas* CHY-1 (PhnI) was unique in that it was able to oxidize at least eight PAHs comprising two to five aromatic rings. Structural studies showed that its catalytic domain featured the largest hydrophobic substrate-binding cavity. The central region of the catalytic pocket is shaped mainly by the side chains of three amino acids, Phe350, Phe404, and Leu356, which contribute to the selection of the shape and form of allowed substrates. The residue Phe350 is conserved in most known RHOs, except for the NBDO and OMO-O$_{86}$. In NBDO, the position of phenylalanine was replaced with isoleucine. Because Phe is larger than Ile, this Phe residue hinders correct positioning relative to the active site and prevents a nitrobenzene molecule from binding to NDO, as it does to NBDO (Friemann et al. 2005). The Phe404 residue also contributes to regiospecificity as does Phe350. The residue is variable among most RHOs so that it may be more important for substrate selection. Another residue, Leu356, enlarges the catalytic pocket of PhnI to be longer, wider, and higher at the entrance than that of NDO-O$_{9816}$ and BPDO-O$_{RHA1}$, because it is replaced by a bulky aromatic residue of Trp or Phe in NDO-O$_{9816}$ and BPDO-O$_{RHA1}$, respectively (Jakoncic et al. 2007b); Fig. 4d and e). Other residues at the entrance of the substrate-binding pocket seem to also exert a greater influence on substrate

Fig. 4 The active site of RHOs with a substrate in the catalytic pocket. (**a**) Wild-type NDO-O$_{9816-4}$ in complex with naphthalene (*in silver*). Seventeen residue side chains were identified to contribute to the overall topology of the substrate cavity mainly by making a hydrophobic surface suitable for interactions with aromatic substrates. Phe 352 was found to be critical in determining the pocket shape. The mononuclear iron (*black sphere*) is also shown. (**b**) Phe-352-Val NDO-O$_{9816-4}$ in complex with phenanthrene (*in silver*). (**c**) Phe-352-Val NDO-O$_{9816-4}$ in complex with anthracene (*in silver*). Val209 and Leu307 are seen "anchoring" the ligand in the active site. (**d**) The catalytic pocket for BPDO-O$_{RHA1}$ complex with biphenyl (*in silver*). The catalytic pocket is very small. These residues Leu323, Ile326, Phe368, and Phe374 are likely to control the orientation/conformation of the bound substrate. (**e**) The catalytic pocket for PhnI complex with benzo[α]pyrene (*in silver*). The catalytic pocket is shaped mainly by the side chains of three amino acids, Phe350, Phe404 and Leu356, which contribute to select the shape and form of allowed substrates. The side chains of Leu 223 and Phe 350 were rotated to let the substrate fit in the pocket. The enzyme–substrate complexes for PhnI, NDO-O$_{9816-4}$, and BPDO-O$_{RHA1}$ were obtained from corresponding structure (PDB access code 2CKF, 1O7G, and 1ULJ, respectively)

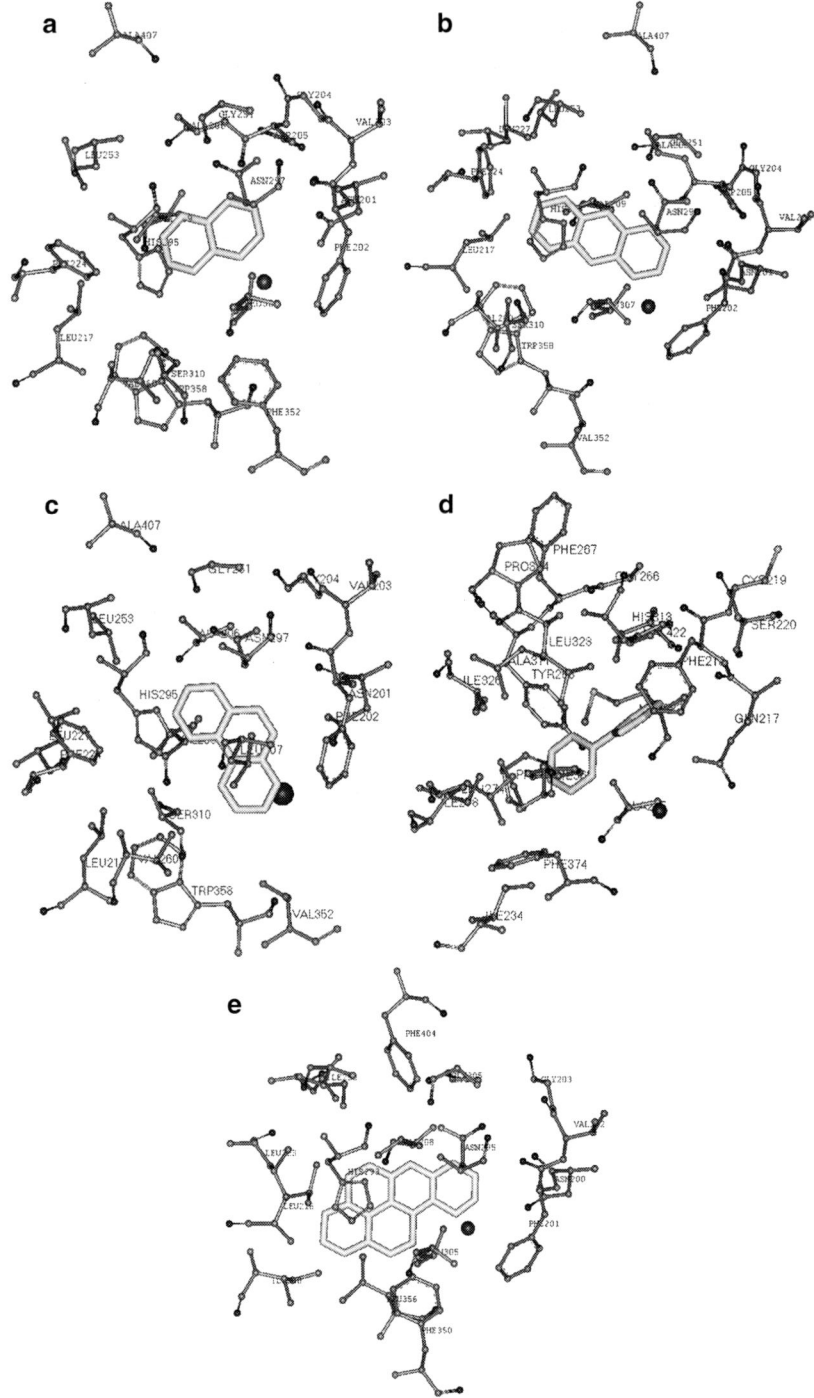

Fig. 4 (continued)

selection. In PhnI, most important are the residues Leu 223 and Leu 226, which are located on loop LI, and the residues Ile 253 and Ile 260, located on loop LII. The diversity in Leu 223 and Ile260 residues must relate to the different substrate specificity observed between members of the naphthalene dioxygenase family (Jakoncic et al. 2007a).

6 Techniques for Improving Ring-hydroxylating Oxygenase Degradation Capabilities

In nature, genes for catabolic functions are considered to have adaptively evolved through various genetic events, resulting in a family of diverse but highly related sequences. The accumulation of small events (e.g., mutations) has likely led to the divergence of RHOs that now demonstrate remarkable differences in substrate specificity. Alternatively, major gene recombinations probably occurred over time and resulted in the evolution of new RHOs that exhibited a broad substrate spectrum. If true, this indicates the possibility of using genetic engineering to produce new RHOs that will have enhanced and expanded degradation capabilities.

Site-directed mutagenesis of a gene is a very important tool in genetic engineering. Deletion, insertion, and point mutations can be directly produced in vitro and the mutant gene can be used to facilitate structure–function relationship studies (Plapp 1995; Peracchi 2001). This method has been broadly used in altering the regio- and enantio-selectivity of RHOs or improving the catalytic efficiency of RHOs. For example, replacement of Phe-352 with smaller amino acids (Gly, Ala, Val, Ile, Leu, Thr) by site-directed mutagenesis resulted in engineering an oxygenase (NDO) that produced significantly more biphenyl *cis*-3,4-dihydrodiol. In addition, the stereochemistry of the biphenyl *cis*-3,4-dihydrodiol was altered. With phenanthrene as a substrate, the A206I F252I and A206I H295I F352I mutants even produced a new product, phenanthrene *cis*-9,10-dihydrodiol (Parales 2003). Compared to the wild type, the F293Q mutant of NBDO-O$_{JS765}$ more rapidly oxidized 2,6-dinitrotoluene (Ju and Parales 2006) by a factor of 2.5. BPDO-O mutants of I335F, T376N, and F377L exhibited altered regiospecificities for various polychlorinated biphenyls (PCBs; Suenaga et al. 2002). Some mutant oxygenases that act on dibenzofuran, dibenzo-p-dioxin, dibenzothiophene, and fluorene were obtained from the KF707 enzyme (Suenaga et al. 1999). Using multiple-site mutagenesis (Peng et al. 2006b), we have made single or multiple substitutions in the residues lining the catalytic pocket of NDO-O$_{9816-4}$ to reconstruct an enzyme that can catalyze a new and extended range of useful reactions (data not shown).

Some experiments have shown that amino acid changes, distant from the active site, can affect substrate specificity. Such changes can work by altering the orientation of active site residues or the conformational dynamics of the entire protein and are therefore difficult to anticipate. DNA shuffling with high-throughput screening is a powerful tool for creating novel enzymes (Xiong et al. 2007a,b). By using the method of DNA shuffling between the BPDO genes of KF707 and LB400,

Kumamaru et al. (1998) have obtained evolved proteins that have improved degradation capabilities for PCBs and biphenyl-related compounds; enhancements also resulted for single aromatic hydrocarbons, such as benzene and toluene, which are poorly attacked by the wild-type enzyme (Kumamaru et al. 1998). Using the same technique, certain evolved proteins capable of recognizing both *ortho*- and *para*-substituted PCBs were also obtained. These variants exhibited superior degradation capabilities toward several tetra- and penta-chlorinated PCBs (Brühlmann and Chen 1999). By using a family shuffling method, a novel BPDO was selected from different BPDO genes that had been amplified from PCB-contaminated soil DNA; this selected BPDO oxygenated 2,2-chlorinated biphenyls onto carbons 5 and 6, which are positions that LB400 BPDO is unable to attack (Veźina et al. 2007).

During the course of evolution, the process of gene combination has produced many proteins that have new or modified functions. Thus, subunit or domain exchange between dioxygenases of different bacterial origins is a good method to generate new proteins with different functions (Bashton and Chothia 2007). The KF707 biphenyl dioxygenase only degrades PCBs through 2,3-dioxygenation. The LB400 biphenyl dioxygenase can metabolize PCBs through both 2,3-dioxygenation and 3,4-dioxygenation, thereby showing a much wider degradation range of PCB congeners. In addition, the KF707 but not LB400 enzyme can oxidize *para*-replaced congeners such as 4,4'-dichlorobiphenyl. On the contrary, the LB400 enzyme but not the KF707 enzyme can oxidize *ortho-meta*-replaced congeners such as 2,5,2',5'-tetrachlorobiphenyl (Kimura et al. 1997). Kimura et al. (1997) constructed a variety of chimeric biphenyl dioxygenase genes by exchanging four common restriction fragments between the KF707 and the LB400 enzyme. It was discovered that a relatively small number of amino acids in the carboxy-terminal half are involved in the recognition of the chlorinated ring and the sites of dioxygenation (Kimura et al. 1997). Hybrid biphenyl dioxygenases, constructed by exchanging subunits, show that both the β and α subunits influence the substrate reactivity pattern toward PCBs. The $\alpha_{LB400}\beta_{B-356}$ hybrid enzyme showed features of both BPDO from LB400 and *C. testosteroni* B-356, wherein the enzyme was able to metabolize 2,2'- and 3,3'-dichlorobiphenyl, and catalyze the 3,4-dioxygenation of 2,5,2',5'-tetrachlorobiphenyl. Similar results were achieved when hybrids were produced between BPDO from Gram-positive *Rhodococcus globerulus* P6 and Gram-negative strains of LB400 or B-356 (Chebrou et al. 1999). The hybrid biphenyl dioxygenases and toluene dioxygenases (TOD), constructed by replacing gene(s) of large and/or small subunits of the terminal dioxygenases, showed the capability of converting various aromatic hydrocarbons. For example, TodC1–BphA2A3A4, TodC1C2–BphA3A4, and BphA1–TodC2–BphA3A4 showed the capability of converting various aromatic hydrocarbons, including benzene, toluene, biphenyl, diphenylmethane, and naphthalene (Furukawa et al. 1994; Hirose et al. 1994). The TodC1–BphA2A3A4 hybrid enzyme is even capable of efficiently degrading trichloroethylene (Furukawa et al. 1994; Suyama et al. 1996).

A method of exchanging the α or β subunit was also used to construct hybrid naphthalene and nitrotoluene dioxygenase enzyme systems to obtain a novel enzyme with a different substrate range, regiospecificity, and catalyzing activity. When the NDO was combined with 2NTDO from the *Acidovorax* sp. strain JS42,

DNTDO from *Burkholderia* sp. strain DNT, or the TDO from *P. putida* F1, active hybrid dioxygenases were obtained for the three combinations of NDO-β2NTDO, NDO-βDNTDO, and DNTDO-β2NTDO (Parales et al. 1998a,b). Hybridizing the electron transport chain of NDO from *Ralstonia* sp. strain U2 with two separate terminal oxygenases, nitrobenzene dioxygenase from *Comamonas* sp. strain JS765, and 2,4-dinitrotoluene dioxygenase from *Burkholderia* sp. strain DNT, several novel enzymes were obtained, which can simultaneously oxidize DNT and naphthalene (Keenan et al. 2004, 2005; Keenan and Wood 2006).

Phytoremediation is a useful technology for removing contaminants from soil, groundwater, and sediment. Aromatic pollutants that have been successfully phytoremediated include benzene (Doty et al. 2007), toluene (Taghavi et al. 2005), PAHs (Aprill and Sims 1990), and PCBs (Mackova et al. 1997). A direct method for enhancing the effectiveness of phytoremediation is to overexpress in transgenic plants the genes responsible for metabolism, uptake, or transport of specific pollutants (Doty 2008; Aken 2009; James and Strand 2009). RHO genes involved in the uptake or detoxification of PCB pollutants have been inserted into such plants to enhance phytoremediation of this important class of pollutants. Each of the three biphenyl dioxygenase components from *Burkholderia xenovorans* LB400 can be produced individually as active proteins, in transgenic tobacco plants. When all expressed proteins were mixed, they can catalyze the oxygenation of 4-chlorobiphenyl to produce detectable amounts of 2,3-dihydro-2,3-dihydroxy-40-chlorobiphenyl (Mohammadi et al. 2007). However, the transgenic line that coordinately expresses these components was not obtained in this work. In our laboratory we are working to create transgenic plants that simultaneously express all components required to produce active naphthalene dioxygenase. To improve the expression of the naphthalene dioxygenase component, the original codons in the target gene are altered according to plant bias codon by PCR, based on a two-step DNA synthesis method (Peng et al. 2006a; Xiong et al. 2004, 2006, 2008).

7 Summary

Numerous aromatic compounds are pollutants to which exposure exists or is possible, and are of concern because they are mutagenic, carcinogenic, or display other toxic characteristics. Depending on the types of dioxygenation reactions of which microorganisms are capable, they utilize ring-hydroxylating oxygenases (RHOs) to initiate the degradation and detoxification of such aromatic compound pollutants. Gene families encoding for RHOs appear to be most common in bacteria.

Oxygenases are important in degrading both natural and synthetic aromatic compounds and are particularly important for their role in degrading toxic pollutants; for this reason, it is useful for environmental scientists and others to understand more of their characteristics and capabilities. It is the purpose of this review to address RHOs and to describe much of their known character, starting with a review as to how RHOs are classified. A comprehensive phylogenetic analysis has revealed that all RHOs are, in some measure, related, presumably by divergent evolution from a common ancestor, and this is reflected in how they are classified. After we describe RHO classification schemes, we address the relationship between RHO structure and function. Structural differences affect substrate specificity and product formation. In the α subunit of the known terminal oxygenase of RHOs, there is a catalytic domain with a mononuclear iron center that serves as a substrate-binding site and a Rieske domain that retains a [2Fe–2S] cluster that acts as an entity of electron transfer for the mononuclear iron center. Oxygen activation and substrate dihydroxylation occurring at the catalytic domain are dependent on the binding of substrate at the active site and the redox state of the Rieske center. The electron transfer from NADH to the catalytic pocket of RHO and catalyzing mechanism of RHOs is depicted in our review and is based on the results of recent studies. Electron transfer involving the RHO system typically involves four steps: NADH-ferredoxin reductase receives two electrons from NADH; ferredoxin binds with NADH-ferredoxin reductase and accepts electron from it; the reduced ferredoxin dissociates from NADH-ferredoxin reductase and shuttles the electron to the Rieske domain of the terminal oxygenase; the Rieske cluster donates electrons to O_2 through the mononuclear iron.

On the basis of crystal structure studies, it has been proposed that the broad specificity of the RHOs results from the large size and specific topology of its hydrophobic substrate-binding pocket. Several amino acids that determine the substrate specificity and enantioselectivity of RHOs have been identified through sequence comparison and site-directed mutagenesis at the active site. Exploiting the crystal structure data and the available active site information, engineered RHO enzymes have been and can be designed to improve their capacity to degrade environmental pollutants. Such attempts to enhance degradation capabilities of RHOs have been made. Dioxygenases have been modified to improve the degradation capacities toward PCBs, PAHs, dioxins, and some other aromatic hydrocarbons. We hope that the results of this review and future research on enhancing RHOs will promote their expanded usage and effectiveness for successfully degrading environmental aromatic pollutants.

Acknowledgments This research was supported by 863 Program (2006AA06Z358; 2006AA10Z117; 2008AA10Z401); Shanghai Key Laboratory and Basic Research Project (07dz22011); and National Natural Science Foundation (06ZR14073).

References

Aken BV (2009) Transgenic plants for phytoremediation: helping nature to clean up environmental pollution. Trends Biotechnol 26:225–227

Aprill W, Sims RC (1990) Evaluation of the use of prairie grasses for stimulating polycyclic aromatic hydrocarbon treatment in soil. Chemosphere 20:253–265

Armengaud J, Timmis KN (1997) Molecular characterization of Fdx1, a putidaredoxin-type [2Fe–2S] ferredoxin able to transfer electrons to the dioxin dioxygenase of *Sphingomonas* sp. RW1. Eur J Biochem 247:833–842

Ashikawa Y, Fujimoto Z, Noguchi H, Habe H, Omori T, Yamane H, Nojiri H (2006) Electron transfer complex formation between oxygenase and ferredoxin components in Rieske nonheme iron oxygenase system. Structure 14:1779–1789

Asturias JA, Diaz E, Timmis KN (1995) The evolutionary relationship of biphenyl dioxygenase from gram-positive *Rhodococcus globerulus* P6 to multicomponent dioxygenases from gram-negative bacteria. Gene 156:11–18

Bashton M, Chothia C (2007) The generation of new protein functions by the combination of domains. Structure 15:85–99

Batie CJ, Ballou DP, Correll CC (1991) Phthalate dioxygenase reductase and related flavin–iron–sulfur containing electron transferases. In: Müller F (ed) Chemistry and biochemistry of flavoenzymes. CRC, Boca Raton, FL, pp 544–556

Brugna M, Nitschke W, Asso M, Guigliarelli B, Lemesle-Meunier D, Schmidt C (1999) Redox components of cytochrome *bc*-type enzymes in acidophilic prokaryotes. II. The Rieske protein of phylogenetically distant acidophilic organisms. J Biol Chem 274:16766–16772

Brühlmann F, Chen W (1999) Tuning biphenyl dioxygenase for extended substrate specificity. Biotechnol Bioeng 63:544–551

Bünz PV, Cook AM (1993) Dibenzofuran 4,4a-dioxygenase from *Sphingomonas* sp. strain RW1: Angular dioxygenation by a three-component enzyme system. J Bacteriol 175: 6467–6475

Butler CS, Mason JR (1997) Structure function analysis of the bacterial aromatic ring-hydroxylating dioxygenases. Adv Microb Physiol 38:47–84

Carredano E, Karlsson A, Kauppi B, Choudhury D Parales RE, Parales JV, Lee K, Gibson DT, Eklund H, Ramaswamy S (2000) Substrate binding site of naphthalene 1,2-dioxygenase: functional implications of indole binding. J Mol Biol 296:701–712

Carrell CJ, Zhang H, Cramer WA, Smith JL (1997) Biological identity and diversity in photosynthesis and respiration: Structure of the lumen-side domain of the chloroplast Rieske protein. Structure 5:1613–1625

Castresana J, Lübben M, Saraste M (1995) New archae bacterial genes coding for redox proteins: implications for the evolution of aerobic metabolism. J Mol Biol 250:202–210

Chang HK, Zylstra GJ (1998) Novel organization of the genes for phthalate degradation from *Burkholderia cepacia* DBO1. J Bacteriol 180:6529–6537

Chebrou H, Hurtubise Y, Barriault D, Sylvestre M (1999) Heterologous expression and characterization of the purified oxygenase component of *Rhodococus globerulus* P6 biphenyl dioxygenase and of chimeras derived from it. J Bacteriol 181:4805–4811

Colbert CL, Couture MMJ, Eltis LD, Bolin J (2000) A cluster exposed: Structure of the Rieske ferredoxin from biphenyl dioxygenase and the redox properties of Rieske Fe–S proteins. Structure 8:1267–1278

Cosper NJ, D. Eby DM, Kounosu A, Kurosawa N, Neidle EL, Kurtz DM (2002) Rieske-type [2Fe–2S] clusters redox-dependent structural changes in archaeal and bacterial. Protein Sci 11:2969–2973

Denke E, Merbitz-zahradnik T, Hatzfeld OM, Snyder CH, Link TA, Trumpower BL (1998) Alteration of the midpoint potential and catalytic activity of the Rieske iron–sulfur protein by changes of amino acids forming hydrogen bonds to the iron–sulfur cluster. J Biol Chem 273:9085–9093

Dong X, Fushinobu S, Fukuda E, Terada T, Nakamura S, Shimizu K, Nojiri H, Omori T, Shoun H, Wakagi T (2005) Crystal structure of the terminal oxygenase component of cumene dioxygenase from *Pseudomonas fluorescens* IP01. J Bacteriol 187:2483–2490

Doty SL, James CA, Moore AL, Vajzovic A, Singleton GL, Ma C, Khan Z, Xin G, Kang JW, Park JY (2007) Enhanced phytoremediation of volatile environmental pollutants with transgenic trees. Proc Natl Acad Sci USA 104:16816–16821

Doty SL (2008) Enhancing phytoremediation through the use of transgenics and endophytes. New Phytologist 179:318–333

Erickson BD, Mondello FJ (1992) Nucleotide sequencing and transcriptional mapping of the genes encoding biphenyl dioxygenase, a multicomponent polychlorinated-biphenyl-degrading enzyme in *Pseudomonas* strain LB400. J Bacteriol 174:2903–2912

Ferraro DJ, Gakhar L, Ramaswamy S (2005) Rieske business: structure–function of Rieske non-heme oxygenases. Biochem Biophys Res Commun 338:175–190

Fontecave M, Ollagnier-de-Choudens S (2008) Iron–sulfur cluster biosynthesis in bacteria: Mechanisms of cluster assembly and transfer. Arch Biochem Biophys 474:226–237

Friemann R, Ivkovic-Jensen MM, Lessner DJ, Gibson CL, Yu DT, Parales RE, Eklund H, Ramaswamy S (2005) Structural insight into the dioxygenation of nitroarene compounds: the crystal structure of nitrobenzene dioxygenase. J Mol Biol 348:1139–1151

Friemann R, Lee K, Brown EN, Gibson DT, klund H, Ramaswamy S (2009) Structures of the multicomponent Rieske non-heme iron toluene 2,3-dioxygenase enzyme system. Acta Crystallog,Sect D 65:24–33

Fukuda M, Yasukochi Y, Kikuchi Y, Nagata Y, Kimbara K, Horiuchi H, Takagi M, Yano K (1994) Identification of the *bphA* and *bphB* genes of *Pseudomonas* sp. strains KKS102 involved in degradation of biphenyl and polychlorinated biphenyls. Biochem Biophys Res Commun 202:850–856

Furukawa K, Hirose J, Hayashida S, Nakamura K (1994) Efficient degradation of trichloroethylene by a hybrid aromatic ring dioxygenase. J Bacteriol 176:2121–2123

Furusawa Y, Nagarajan V, Tanokura M, Masai E, Fukuda M, Senda T (2004) Crystal structure of the terminal oxygenase component of biphenyl dioxygenase derived from *Rhodococcus* sp. strain RHA1. J Mol Biol 342:1041–1052

Gakhar L, Malik ZA, Allen CC, Lipscomb DA, Larkin MJ, Ramaswamy S (2005) Structure and increased thermostability of *Rhodococcus* sp. naphthalene 1,2-dioxygenase. J Bacteriol 187:7222–7231

Gibson DT (1971) The microbial oxidation of aromatic compounds. Crit Rev Microbiol 1:199–223

Gibson DT, Resnick SM, Lee K, Brand JM, Torok DS, Wackett LP, Schocken MJ, Haigler BE (1995) Desaturation, dioxygenation, and monooxygenation reactions catalyzed by naphthalene dioxygenase from *Pseudomonas* sp. strain 9816-4. J Bacteriol 177:2615–2621

Guergova-Kuras M, Kuras R, Ugulava N, Hadad I, Crofts AR (2000) Specific mutagenesis of the Rieske iron–sulfur protein in *Rhodobacter sphaeroides* shows that both the thermodynamic gradient and the pK of the oxidized form determine the rate of quinol oxidation by the bc1 complex. Biochemistry 39:7436–7444

Harayama S, Rekik M, Bairoch A, Neidle EL, Ornston LN (1991) Potential DNA slippage structures acquired during evolutionary divergence of *Acinetobacter calcoaceticus* chromosomal benABC and *Pseudomonas putida* TOL pWW0 plasmid xylXYZ, genes encoding benzoate dioxygenases. J Bacteriol 173:7540–7548

Hirose J, Suyama A, Hayashida S, Furukawa K (1994) Construction of hybrid biphenyl (bph) and toluene (tod) genes for functional analysis of aromatic ring dioxygenase. Gene 138:27–33

Irie S, Doi S, Yorifuji T, Takagi M, Yano K (1987) Nucleotide sequencing and characterization of the genes encoding benzene oxidation enzymes of *Pseudomonas putida*. J Bacteriol 169: 5174–5179

Jakoncic J, Jouanneau Y, Meyer C, Stojanoff V (2007a) The crystal structure of the ring-hydroxylating dioxygenase from *Sphingomonas* CHY-1. FEBS J 274:2470–2481

Jakoncic J, Jouanneau Y, Meyer C, Stojanoff V (2007b) The catalytic pocket of the ring-hydroxylating dioxygenase from *Sphingomonas* CHY-1. Biochem Biophy Res Commu 352:861–866

James CA, Strand S (2009) Phytoremediation of small organic contaminants using transgenic plants. Curr Opin Biotechnol 20:237–241

Jeffrey AM, Yeh HJ, Jerina DM, Patel TR, Davey JF, Gibson DT (1975) Initial reactions in the oxidation of naphthalene by *Pseudomonas putida*. Biochemistry 14:575–584

Jiang H, Parales RE, Lynch NA, Gibson DT (1996) Site-directed mutagenesis of conserved amino acids in the alpha-subunit of toluene dioxygenase: potential mononuclear nonheme iron coordination sites. J Bacteriol 178:3133–3139

Jones RM, Britt-Compton B, Williams PA (2003) The naphthalene catabolic (*nag*) genes of *Ralstonia* sp. strain U2 are an operon that is regulated by NagR, a LysR-type transcriptional regulator. J Bacteriol 185:5847—5853

Ju KS, Parales RE (2006) Control of substrate specificity by active site residues in nitrobenzene dioxygenase. Appl Environ Microbiol 72:1817–1824

Junker F, Kiewitz R, Cook AM (1997) Characterization of the *p*-toluenesulfonate operon tsaMBCD and tsaR in *Comamonas testosteroni* T-2. J Bacteriol 179:919–927

Karlsson A, Parales JV, Parales RE, Gibson DT, Eklund H, Ramaswamy S (2003) Crystal structure of naphthalene dioxygenase: side-on binding of dioxygen to iron. Science 299: 1039–1042

Kauppi B, Lee K, Carredano E, Parales RE, Gibson DT, Eklund H, Ramaswamy S (1998) Structure of an aromatic-ring-hydroxylating dioxygenase-naphthalene 1.2-Dioxygenase. Structure 6:571–586

Keenan BG, Leungsakul T, Smets BF, Wood TK (2004) Saturation mutagenesis of *Burkholderia cepacia* R34 2,4-dinitrotoluene dioxygenase at DntAc valine 350 for synthesizing nitrohydroquinone, methylhydroquinone, and methoxyhydroquinone. Appl Environ Microbiol 70:3221–3222

Keenan BG, Leungsakul T, Smets BF, Mori M, Henderson DE, Wood TK (2005) Protein engineering of the archetypal nitroarene dioxygenase of *Ralstonia* sp. strain U2 for activity on aminonitrotoluenes and dinitrotoluenes through alpha-subunit residues leucine 225, phenylalanine 350, and glycine 407. J Bacteriol 187:3302–3310

Keenan BG, Wood TK (2006) Orthric Rieske dioxygenases for degrading mixtures of 2,4-dinitrotoluene/naphthalene and 2-amino-4, 6-dinitrotoluene/4-amino-2,6-dinitrotoluene. Appl Microbiol Biotechnol 73:827–838

Kimura N, Nishi A, Goto M, Furukawa K (1997) Functional analyses of a variety of chimeric dioxygenases constructed from two biphenyl dioxygenases that are similar structurally but different functionally. J Bacteriol 179:3936–3943

Kolling DJ, Brunzelle JS, Lhee SM, Crofts AR, Nair SK (2007) Atomic resolution structures of Rieske iron–sulfur protein: role of hydrogen bonds in tuning the redox potential of iron–sulfur clusters. Structure 15:29–38

Kovaleva EG, Neibergall MB, Chakrabarty S, Lipscomb JD (2007) Finding intermediates in the O_2 activation pathways of non-heme iron oxygenases. Acc Chem Res 40:475–483

Kumamaru T, Suenaga H, Mitsuoka M, Watanabe T, Furukawa K (1998) Enhanced degradation of polychlorinated biphenyls by directed evolution of biphenyl dioxygenase. Nat Biotechnol 16:663–666

Kurkela S, Lehväslaiho H, Palva ET, Teeri TH (1988) Cloning, nucleotide sequence and characterization of genes encoding naphthalene dioxygenase of *Pseudomonas putida* strain NCIB9816. Gene 73:355–362

Kweon O, Kim SJ, Baek S, Chae JC, Adjei MD, Baek DH, Kim YC, Cerniglia CE (2008) A new classification system for bacterial Rieske non-heme iron aromatic ring-hydroxylating oxygenases. BMC Biochemistry 9:11.

Lee KS, Parales JV, Friemann R, Parales RE (2005) Active site residues controlling substrate specificity in 2-nitrotoluene dioxygenase from *Acidovorax* sp. strain JS42. J Ind Microbiol Biotechnol 32:465–473

Mackova M, Macek T, Ocenaskova J, Burkhard J, Demnerova K, Pazlarova J (1997) Biodegradation of polychlorinated biphenyls by plant cells. Int Biodeterior Biodegrad 39: 317–25

Mason JR, Cammack R (1992) The electron-transport proteins of hydroxylating bacterial dioxygenases. Ann Rev Microbio 46:277–305

Martin VJ, Mohn WW (1999) A novel aromatic-ring-hydroxylating dioxygenase from the diterpenoid degrading bacterium *Pseudomonas abietaniphila* BKME-9. J Bacteriol 181:2675–2682

Martins BM, Svetlitchnaia T, Dobbek H (2005) 2-Oxoquinoline 8-monooxygenase oxygenase component: active site modulation by Rieske-[2Fe-2S] center oxidation/reduction. Structure 13:817–824

McMurry JE (2004) Organic chemistry, 6th edn. Brooks/Cole, Pacific Grove, CA

Merbitz-Zahradnik T, Zwicker K, Nett JH, Link TA, Trumpower BL (2003) Elimination of the disulfide bridge in the Rieske iron–sulfur protein allows assembly of the [2Fe–2S] cluster into the Rieske protein but damages the ubiquinol oxidation site in the cytochrome bc1 complex, Biochemistry 42:13637–13645

Mohammadi M, Chalavi V, Novakova-Sura M, Laliberte JF, Sylvestre M (2007) Expression of bacterial biphenyl–chlorophenyl dioxygenase genes in tobacco plants. Biotechnol Bioeng 97: 496–505

Nakatsu CH, Straus NA, Wyndham RC (1995) The nucleotide sequence of the Tn5271 3-chlorobenzoate 3, 4-dioxygenase genes (cbaAB) unites the class IA oxygenases in a single lineage. Microbiology 141:485–495

Nam JW, Nojiri H, Yoshida T, Habe H, Yamane H, Omori T (2001) New classification system for oxygenase components involved in ring-hydroxylating oxygenations. Biosci Biotechnol Biochem 65:254–263

Nam JW, Nojiri H, Noguchi H, Uchimura H, Yoshida T, Habe H, Yamane H, Omori T (2002) Purification and characterization of carbazole 1,9a-dioxygenase, a three-component dioxygenase system of *Pseudomonas resinovorans* strain CA10. Appl Environ Microbiol 68:5882–5890

Nam JW, Noguchi H, Fujimoto Z, Mizuno H, Ashikawa Y, Abo M. Fushinobu S, Kobashi N, Wakagi T, Iwata K, Yoshida T, Habe H, Yamane H, Omori T, Nojiri H (2005) Crystal structure of the ferredoxin component of carbazole 1,9a-dioxygenase of *Pseudomonas resinovorans* strain CA10, a novel Rieske non-heme iron oxygenase system. Proteins Struct Funct Genet 58:779–789

Neibergall MB, Stubna A, Mekmouche Y, Münck E, Lipscomb JD (2007) Hydrogen peroxide dependent *cis*-dihydroxylation of benzoate by fully oxidized benzoate 1,2-dioxygenase. Biochemistry 46:8004–8016

Neidle EL, Harnett C, Ornston LN, Bairoch A, Rekik M, Hara-yama S (1991) Nucleotide sequence of the *Acinetobacter calcoaceticus* benABC genes for benzoate 1, 2-dioxygenase reveal evolutionary relationships among multicomponent oxygenases. J Bacteriol 173: 5385–5395

Nojiri H, Ashikawa Y, Noguchi H, Nam JW, Urata M, Fujimoto Z, Uchimura H, Terada T, Nakamura S, Shimizu K, Yoshida T, Habe H, Omori T (2005) Structure of the terminal oxygenase component of angular dioxygenase, carbazole 1,9a-dioxygenase. J Mol Biol 351:355–370

Parales JV, Parales RE, Resnick SM, Gibson DT (1998a) Enzyme specificity of 2-nitrotoluene 2,3-dioxygenase from Pseudomonas sp. strain JS42 is determined by the C-terminal region of the alpha-subunit of the oxygenase component. J Bacteriol 180:1194–1199

Parales RE, Emig MD, Lynch NA, Gibson DT (1998b) Substrate specificities of hybrid naphthalene and 2,4-dinitrotoluene dioxygenase enzyme systems. J Bacteriol 180:2337–2344

Parales RE, Parales JV, Gibson DT (1999) Aspartate 205 in the catalytic domain of naphthalene dioxygenase is essential for activity. J Bacteriol 181:1831–1837

Parales RE, Lee K, Resnick SM, Jiang H, Lessner DJ, Gibson DT (2000) Substrate specificity of naphthalene dioxygenase: effect of specific amino acids at the active site of the enzyme. J Bacteriol 182:16417–1649

Parales RE (2003) The role of active-site residues in naphthalene dioxygenase. J Ind Microbiol Biotechnol 30:271–278

Peng RH, Yao QH, Xiong AS, Cheng ZM, Li Y (2006a) Codon modifications and an endoplasmic reticulum-targeting sequence additively enhance expression of an *Aspergillus* phytase gene in transgenic canola. Plant Cell Rep 25:124–132

Peng RH, Xiong AS, Yao QH (2006b) A direct and efficient PAGE-mediated overlap extension PCR method for gene multiple-site mutagenesis. Appl Microbiol Biotechnol 73:234–240

Peng RH, Xiong AS, Xue Y, Fu XY, Gao F, Zhao W, Tian YS, Yao QH (2008) Microbial biodegradation of polyaromatic hydrocarbons. FEMS Microbiol Rev 32:927–955

Peracchi A (2001) Enzyme catalysis: removing chemically "essential" residues by site-directed mutagenesis. Trends Biochem Sci 26:497–503

Pieper DH, Rieneke W (2001) Engineering bacteria for bioremediation. Curr Opin Biotechnol 11:262–270

Plapp BV (1995) Site-directed mutagenesis: a tool for studying enzyme catalysis. Methods Enzymol 249:91–119

Raag R, Poulos TL (1989) Crystal structure of the carbon monoxide–substrate–cytochrome P-450CAM ternary complex. Biochemistry 28:7586–7592

Resnick SM, Lee K, Gibson DT (1996) Diverse reactions catalyzed by naphthalene dioxygenase from *Pseudomonas* sp. strain NCIB 9816. J Ind Microbiol 17:438–457

Reineke W, Knackmuss HJ (1988) Microbial degradation of haloaromatics. Ann Rev Microbiol 42:263–287

Rieske JS, Maclennan DH, Coleman R (1964) Isolation and properties of an iron–protein from the (reduced coenzyme Q) –cytochrome *c* reductase complex of respiratory chain. Biochem Biophys Res Commun 15:338–344

Romine MF, Stillwell LC, Wong KK, Thurston SJ, Sisk EC, Sensen C, Gaasterland T, Fredrickson JK, Saffer JD (1999) Complete sequence of a 184-kilobase catabolic plasmid from *Sphingomonas aromaticivorans* F199. J Bacteriol 181:1585–1602

Rosche B, Tshisuaka B, Fetzner S, Lingens F (1995) 2-Oxo-1,2-dihydroquinoline 8-monooxygenase, a two-component enzyme system from *Pseudomonas putida* 86. J Boil Chem 270:17836–17842

Rosche B, Tshisuaka B, Hauer B, Lingens F, Fetzner S (1997) 2-Oxo-1,2-dihydroquinoline 8-monooxygenase: phylogenetic relationship to other multicomponent nonheme iron oxygenases. J Bacteriol 179:3549–3554

Sato S, Nam JW, Kasuga K, Nojiri H, Yamane H, Omori T (1997) Identification and characterization of gene encoding carbazole 1,9a-dioxygenase in *Pseudomonas* sp. strain CA10. J Bacteriol 179:4850–4858

Schäfer G, Purschke W, Schmidt CL (1996) On the origin of respiration: electron transport proteins from archaea to man. FEMS Microbiol Rev 18:173–188

Schmidt CL, Shaw L (2001) A comprehensive phylogenetic analysis of Rieske and Rieske-type iron–sulfur proteins. J Bioeng Biom 33:9–26

Schröter T, Hatzfeld OM, Gemeinhardt S, Korn M, Friedrich T, Ludwig B, Link TA (1998) Mutational analysis of residues forming hydrogen bonds in the Rieske [2Fe–2S] cluster of the cytochrome *bc*1 complex in *Paracoccus denitrificans*. Eur J Biochem 255:100–106

Senda M, Kishigami S, Kimura S, Fukuda M, Ishida T, Senda T (2007) Molecular mechanism of the redox-dependent interaction between NADH dependent ferredoxin reductase and Rieske-type [2Fe–2S] ferredoxin. J Mol Biol 373:382–400

Simon MJ, Osslund TD, Saunders R, Ensley BD, Suggs S, Harcourt A, Suen WC, Cruden DL, Gibson DT, Zylstra GJ (1993) Sequences of genes encoding naphthalene dioxygenase in *Pseudomonas putida* strains G7 and NCIB 9816-4. Gene 127:31–37

Stingley RL, Khan AA, Cerniglia CE (2004) Molecular characterization of a phenanthrene degradation pathway in *Mycobacterium vanbaalenii* PYR-1. Biochem Biophys Res Commun, 322:133–146

Subramanian V, Liu TN, Yeh WK, Gibson DT (1979) Toluene dioxygenase: purification of an iron–sulfur protein by affinity chromatography. Biochem Biophys Res Commun 91:1131–1139

Suenaga H, Nishi A, Watanabe T, Sakai M, Furukawa K (1999) Engineering a hybrid pseudomonad to acquire 3,4-dioxygenase activity for polychlorinated biphenyls. J Biosci Bioeng 87: 430–435

Suenaga H, Watanabe T, Sato M, Sakai M, Ngadiman M, Furukawa K (2002) Alteration of regiospecificity in biphenyl dioxygenase by active-site engineering. J Bacteriol 184:3682–3688

Suyama A, Iwakiri R, Kimura N, Nishi A, Nakamura K, Furukawa K (1996) Engineering hybrid pseudomonads capable of utilizing a wide range of aromatic hydrocarbons and of efficient degradation of trichloroethylene. J Bacteriol 178:4039–4046

Taghavi S, Barac T, Greenberg B, Borremans B, Vangronsveld J, van der Lelie D (2005) Horizontal gene transfer to endogenous endophytic bacteria from poplar improves phytoremediation of toluene. Appl Environ Microbiol 71: 8500–8505

Takizawa N, Kaida N, Torigoe S, Moritani T, Sawada T, Satoh S, Kiyohara H (1994) Identification and characterization of genes encoding polycyclic aromatic hydrocarbon dioxygenase and polycyclic aromatic hydrocarbon dihydrodiol dehydrogenase in *Pseudomonas putida* OUS82. J Bacteriol 176:2444–2449

Tarasev M, Ballou DP (2005) Chemistry of the catalytic conversion of phthalate into its *cis*-dihydrodiol during the reaction of oxygen with the reduced form of phthalate dioxygenase. Biochemistry 44:6197–6207

Treadway SL, Yanagimachi KS, Lankenau E Lessard PA, Stephanopoulos G, Sinskey AJ (1999) Isolation and characterization of indene bioconversion genes from Rhodococcus strain I24. Appl Microbiol Biotechnol 51:786–793

Trumpower BL (1981) Function of the iron–sulfur protein of the cytochrome bc1 segment in electron-transfer and energy-conserving reactions of the mitochondrial respiratory chain. Biochim Biophys Acta 639:129–155

Trumpower BL, Gennis RB (1994) Energy transduction by cytochrome complexes in to transmembrane proton translocation. Ann Rev Biochem 63:675–716

Ugulava NB, Crofts AR (1998) CD-monitored redox titration of the Rieske Fe–S protein of *Rhodobacter sphaeroides*: pH dependence of the midpoint potential in isolated bc1 complex and in membranes. FEBS Lett 440:409–413

Vaillancourt FH, Bolin JT, Eltis LD (2006) The ins and outs of ring-cleaving dioxygenases. Crit Rev Biochem Mol Bio 41: 241–267

Vežina J, Barriault D, Sylvestre M (2007) Family shuffling of soil DNA to change the regiospecificity of *Burkholderia xenovorans* LB400 biphenyl dioxygenase. J Bacteriol 189:779–788

Wolfe MD, Parales JV, Gibson DT, Lipscomb JD (2001) Single turnover chemistry and regulation of O_2 activation by the oxygenase component of naphthalene 1,2-dioxygenase. J Biol Chem 276:1945—1953

Xiong AS, Yao QH, Peng RH, Li X, Fan HQ, Cheng ZM, Li Y (2004) A simple, rapid, high-fidelity and cost-effective PCR based two-step DNA synthesis method for long gene sequences. Nucleic Acids Res 32: e98

Xiong AS, Yao QH, Peng RH, Duan H, Li X, Fan HQ, Cheng ZM, Li Y (2006) PCR-based accurate synthesis of long DNA sequences. Nat Protoc 1: 791–797

Xiong AS, Peng RH, Zhuang J, Liu JG, Gao F, Fang X, Cai B, Yao QH (2007a) A semi-rational design strategy of directed evolution combined with chemical synthesis of DNA sequences. Biol Chem 388: 1291–1300

Xiong AS, Peng RH, Zhuang J, Li X, Xue Y, Liu JG, Cai B, Chen JM, Yao QH (2007b) Directed evolution of a beta-galactosidase from *Pyrococcus woesei* resulting in increased thermostable beta-glucuronidase activity. Appl Microbiol Biotechnol 77:569–578

Xiong AS, Peng RH, Zhuang J, Gao F,Li Y, Cheng ZM, Yao QH (2008) Chemical gene synthesis: strategies, softwares, error corrections, and applications. FEMS Microbiol Rev 32: 522–540

Ziffer H, Jerina DM, Gibson DT, Kobal VM (1973) Absolute stereochemistry of the (+)-*cis*-1, 2-dihydroxy-3-methylcyclohexa-3,5-diene produced from toluene by *Pseudomonas putida*. J Am Chem Soc 95:4048–4049

Environmental Implications of Oil Spills from Shipping Accidents

Justyna Rogowska and Jacek Namieśnik

Contents

1 Introduction 95
2 Oil in the Marine Environment 98
3 Fate and Behavior of Oil Spills in the Marine Environment 100
4 The Impact of Oil Spills on Seabirds, Fish, and Marine Animals 104
5 Shipping Accidents in the Baltic Sea 106
6 Implications of Shipping-oil Spills 109
7 Summary 111
References 111

1 Introduction

It was only during the second half of the twentieth century that people became aware of the negative effects of environmental pollution. For many years, the prevailing opinion had been that the vast and continuously flowing masses of water in the seas and oceans would absorb all human-derived pollutants and wastes dumped into that environment, and that these pollutants would not cause any permanent negative environmental damage (Łopuski 1974). But reports from U Thant (United Nations (UN) Secretary General) in 1969 and the Club of Rome (Meadows et al. 1973) inspired a change of attitude, which raised concerns about the environment and its protection. Since that time, marine pollution has grown into a topic of global significance.

Pollution of sea water by petroleum products began toward the end of the nineteenth century as sailing vessels were replaced by steamships (Camphuysen 2007). The increased reliance on oil during the twentieth century, particularly in motor vehicles and as an industrial energy source, made this raw material both

J. Namieśnik (✉)
Department of Analytical Chemistry, Chemical Faculty, Gdańsk University of Technology, ul, Narutowicza 11/12, 80-233 Gdańsk, Poland
e-mail: chemanal@pg.gda.pl

economically and politically important. In 2007, the total world oil production was 85 million barrels per day (mbpd) (E.I.A. 2009); approximately half of this volume of oil (43 mbpd) was transported along the international shipping lanes of the world's seas. In Table 1, we present the total consumption of petroleum products at intervals over recent decades.

Table 1 Total consumption of petroleum products in recent decades (Barrels Per Day) (E.I.A. 2009)

1980	1990	2000	2008
63,113,271	66,686,600	76,741,281	85,777,270

The growing demand for oil products has resulted in an ongoing search for new oilfields on land and at sea and for increasingly cheaper and faster methods of getting petroleum products to the markets. Accordingly, ever larger tankers to transport the oil are being designed and constructed, and the tanker traffic along international sea lanes is growing progressively heavier. The first of the sea-going supertankers began service in the early 1950s. In 1959, the first tanker with a capacity of >100,000 t was launched, and the largest tanker ever – the *Jahre Viking* (564,763 dead weight tons (DWT)) – began service in 1979 (Wiewióra et al. 2007). It is estimated that from 1914 to 1953 the marine fluid-fuel shipping tonnage increased from 1.51 to 81 million brutto register tons (BRT) (Borakowski 1955). Of the 71,929 vessels in service on the high seas around the world in 2007, more than 11,000 were tankers dedicated to the transport of oil and chemicals, with a combined gross tonnage (GT) of 232,757,000. Almost 10% of all tankers had a GT of 60,000 each, and 16% had a GT between 25,000 and 60,000 each (EMSA (The European Maritime Safety Agency) 2009).

Although sea transport is among the most environmentally friendly ways of moving goods around, it does have negative effects. These include

1. Regular ship operation – wastes produced and discharged into the water/atmosphere, cleaning of oil tanks, reloading in harbors, etc.
2. Exceptional events – collisions, accidents, ship groundings, accidental leaks.

A total of 4,999 oil spills were reported between 1974 and 2008 (according to The International Tanker Owners Pollution Federation Limited (ITOPF) database), resulting from operations such as oil loading, discharging, or bunkering; only 0.6% of these were large spills (>700 t). Collisions and groundings were the most important causes of such spills (ITOPF 2009). The total number of oil spills resulting from such accidents, during the period 1974–2008, was 4,369, of which >7% was classified as large spills (>700 t), 18% as medium spills (7–700 t), and almost 75% as small spills (<7 t; ITOPF 2009). Figure 1a and b shows the distribution of causes that have resulted in oil spills (ITOPF 2009).

Although the main source of marine oil pollution derives from ship operation, the greatest threat results from tanker accidents or the malfunctioning of other vessels and off-shore drilling rigs. Every accident involving an oil tanker may pollute both

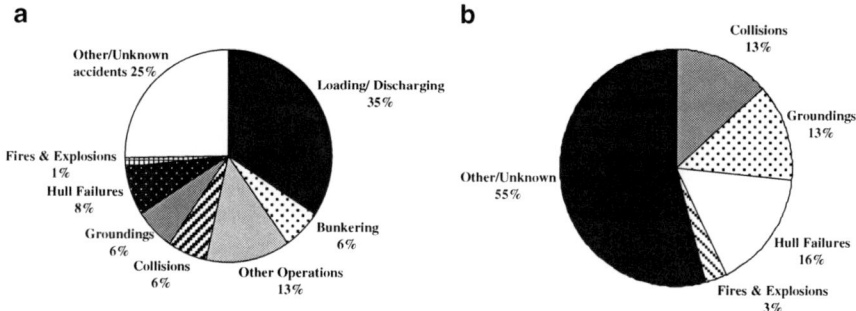

Fig. 1 (**a**) General causes of oil spills worldwide from 1974 to 2008 (The International Tanker Owners Pollution Federation Ltd (ITOPF) 2009); (**b**) causes of accidental oil spills from 1974 to 2008 (ITOPF 2009)

waters and adjacent shore areas (Bądkowski 1977). The amount of oil leaking from a vessel after an accident may be as high as several hundred thousand tons. Table 2 shows a list of the accidents that have led to the largest (>50,000 t) oil spills between 1960 and 2002 (GESAMP (Joint Group of Experts on the Scientific Aspects of Marine) 2007).

The first accident that brought home to the international community the extreme seriousness of oil spills at sea was the disaster involving the Liberian tanker *Torrey Canyon*. On March 18, 1967, she ran onto rocks off the coast of Cornwall, England. As a result, over 120,000 t of crude oil was spilled into the ocean. Never before had a tanker disaster at sea left such devastating repercussions, either for the sea itself or for the adjacent shores (Pietraszek 1967). The *Torrey Canyon* event was an environmental disaster for many organisms of the same magnitude that the sinking of the *Titanic* was for humans. Oil pollution caused by this event extended for a distance of 120 miles along the Cornish coast and 80 km along the French coast on the opposite side of the English Channel. The resident populations of marine organisms and seabirds of these shores suffered unimaginable losses. It is estimated that the ecosystem took 6 years to recover. But the most serious tanker accident ever to have occurred in European seas was the one involving the *Amoco Cadiz* on March 17, 1978, on the coast of Brittany, France. More than 230,000 t of oil flowed into the sea as a result of this accident.

Although tanker disasters are thankfully very rare events, they frequently have dramatic consequences for the marine environment when they do occur. The main deleterious effects of shipping accidents are as follows:

- pollution of the sea water (in most cases, a short-term effect);
- pollution of the seabed (including possible effects on spawning grounds);
- pollution of the atmosphere (especially if the spilled oil catches fire);
- mortality or morbidity of seabirds and mammals;
- pollution of shores (both recreational areas and wildlife habitats can be seriously affected) (HELCOM (Helsinki Commission) 2006).

Table 2 Worldwide tanker oil spills (>50,000 t of oil) between 1960 and 2002 (GESAMP 2007)

Year	Name of tanker	Location	Oil lost (t)
1983	*Castillo de Bellver*	South Africa	267,007
1978	*Amoco Cadiz*	France	233,565
1988	*Odyssey*	North Atlantic, Canada	146,599
1979	*Atlantic Empress2*	Trinidad and Tobago	145,252
1991	*Haven*	Italy	144,000
1979	*Atlantic Empress2*	Barbados	141,102
1967	*Torrey Canyon*	United Kingdom	129,857
1972	*Sea Star*	Oman	128,891
1980	*Irenes Serenade*	Greece	124,490
1971	*Texaco Denmark*	Belgium	107,143
1979	*Independentza*	Turkey	98,255
1969	*Julius Schindler*	Portugal	96,429
1976	*Urquiola*	Spain	95,714
1993	*Braer*	United Kingdom	85,034
1975	*Jakob Maersk*	Portugal	82,503
1992	*Aegean Sea*	Spain	74,490
1985	*Nova*	Iran	72,626
1996	*Sea Empress*	United Kingdom	72,361
1989	*Khark 5*	Morocco	70,068
1971	*Wafra*	South Africa	68,571
2002	*Prestige*	Spain	63,000
1960	*Sinclair Petrolore*	Brazil	60,000
1983	*Assimi*	Oman	53,741
1974	*Yuyo Maru No. 10*	Japan	53,571
1971	*ABT Summer*	Angola	51,020
1992	*Katina P.*	South Africa	51,020
1964	*Heimvard*	Japan	50,000

2 Oil in the Marine Environment

Petroleum is a complex mixture of thousands of different organic compounds, formed from a variety of organic materials that are chemically converted under differing geological conditions over long periods of time. Crude oils contain primarily carbon and hydrogen (which form a wide range of hydrocarbons from light gases to heavy residues), but also contain smaller amounts of sulfur, oxygen, and nitrogen as well as metals such as nickel, vanadium, and iron (Wang et al. 1999). The principal constituents of crude oil include

- cyclic hydrocarbons, e.g., cyclohexane;
- aliphatic hydrocarbons, e.g., n-butane, isobutane, n-hexane;
- aromatic hydrocarbons, e.g., toluene, m-xylene, styrene, benzo(a)pyrene; and
- other elements, e.g., sulfur, nitrogen, metals.

The behavior of hydrocarbons in the marine environment depends on their density and solubility in water. The average density of the petroleum fraction ranges from <0.878 g/cm^3 for light fractions to >0.884 g/cm^3 for heavy fractions (Surygała 2001). The average density of sea water is estimated to be 1.025 g/cm^3. This means that the lighter oil fractions rise to the water surface after the volatile constituents have evaporated. The heavy fractions sink to the seabed from where some constituents may return to the surface after having been subjected to biological processes on the seabed (Różańska 1987). The density of crude oil depends on the chemical composition of the constituent petroleum products: the density is higher with an elevated content of aromatic hydrocarbons, but lower if paraffin hydrocarbons predominate (Surygała 2001). The solubility of petroleum products in water is an important factor that affects their mobility and distribution within the environment. Mobility and distribution is affected by the adsorption and desorption of oil components on sediments and their volatility and migration into air from aquatic systems. Moreover, the chemical components of oil may be transformed in water by several processes, including hydrolysis, photolysis, oxidation, reduction, and biodegradation (Page et al. 2000).

Hydrocarbons vary greatly in their water solubility. Solubility depends on structure and usually decreases with increasing molecular weight. For example, the solubility of benzene is 0.022 mol/L, but that of tetradecane is only 1.1×10^{-8} mol/L (Page et al. 2000). The rate and extent to which oil dissolves in water depend upon its composition, the water temperature, how rapidly the oil spreads on the water surface, the degree of water turbulence, and degree of oil dispersion in the water. The heavier components of crude oil are virtually insoluble in sea water, whereas lighter compounds, particularly aromatic hydrocarbons such as benzene and toluene, are slightly soluble. However, these lighter compounds are also the most volatile and are lost very rapidly by evaporation, typically 10–1,000 times faster than by dissolution. Thus, concentrations of dissolved hydrocarbons in sea water rarely exceed 1.0 ppm. Therefore, dissolution makes no significant contribution to the removal of oil from the surface of the sea (ITOPF 2002).

Some hydrocarbons, mainly low molecular weight compounds, which are relatively toxic, do dissolve in sea water. However, the extent of this dissolution is small, i.e., <1% of spilled oil. The dissolved fraction quickly becomes diluted and soon degrades (Kingston 2002). For example, the *Prestige*, a tanker flying the Bahamian flag, was carrying heavy fuel oil (M-100 type; viscosity at 15°C = 100,000 cSt; density at 15°C = 0.992 kg/L; sulfur content = 2.28%; aliphatic hydrocarbon content = 22%; aromatic hydrocarbon content = 50%; resin and asphaltene content = 28%) when it was involved in an oil spill. Once in the water, this oil became strongly emulsified and formed patches, greatly complicating the recovery operation (Albaigés et al. 2006). The *Baltic Carrier* tanker, carrying the same type of heavy fuel as the *Prestige*, ran aground on the Danish coast in April 2001. In this second accident, the high viscosity of the oil leaking out of the vessel prevented its being contained and skimmed from the open sea surface with the available pollution control equipment on board the environmental protection vessels. This highly viscous oil was equally difficult to recover from shallow, coastal waters (HELCOM SEA 2002).

3 Fate and Behavior of Oil Spills in the Marine Environment

Two processes determine the movement behavior of oil spilled onto the sea surface: advection and spreading (National Research Council 1985). Typically, spilled oil spreads over the surface of the water, forming a film (Kachel 2008). Slicks are roughly classified as either "thin" ones, <10 μm thick, and "thick" ones, often millimeters or even centimeters thick (National Research Council 1985). Spreading is governed by gravity, viscous forces, surface tension, and the balance among them (Fay's theory), whereas advection depends on the activity of winds and sea currents. On January 16, 2001, the Ecuador-registered tanker *Jessica* ran aground in Wreck Bay on San Cristobal Island in the southeast of the Galapagos Archipelago, spilling its cargo of some 600 t of diesel and 300 t of bunker fuel into the sea (Gelin et al. 2003). It was the largest oil spill ever to have taken place on the Galapagos Islands and had the potential to cause irreparable damage to the vulnerable marine fauna that are unique to the islands. Fortunately, wind and current action drew the bulk of the spilled oil away from the shores of San Cristobal, and the mix of bunker fuel and the lighter diesel oil resulted in its rapid and wide dispersion (Kingston et al. 2003). In contrast, with the *Exxon Valdez* disaster, winds of more than 70 mph, on the third day after the grounding, rendered containment of the oil on the water impossible. The result was that the shoreline habitat in the southwestern segment of Prince William Sound was impacted. During the subsequent few weeks more than 1100 miles of shoreline in south-central Alaska were affected by oil to varying degrees (Maki 1991). In a third example, the oil slick from the wreck of the *Prestige* drifted about on the sea surface for almost 1 year (after November 2002), finally making a distant landfall, between May and August 2003 (Dies et al. 2007).

The fate of spilled oil in the marine environment depends on the physical and chemical properties of oil, the characteristics of the environment affected, as well as the physical, chemical, and biological processes occurring there, such as evaporation, dispersion, microbial degradation, photo-oxidation, and interaction between oil and sediments (Wang et al. 1999). The combination of these processes, known as "weathering," reduces the concentrations of hydrocarbons in sediments and water and alters the chemical composition of spilled oils. Figure 2 illustrates the kinds of processes that occur in the marine environment following an oil spill.

For many oil spills, the most important process that affects the mass balance of spilled oil is evaporation. The rate of this process depends primarily on ambient temperature, wind speed, and oil composition. Evaporation is at its most intensive in the first hours and days after the spill. Within a few days following a spill, light crude oils may lose up to 75% of their initial volume and medium crudes up to

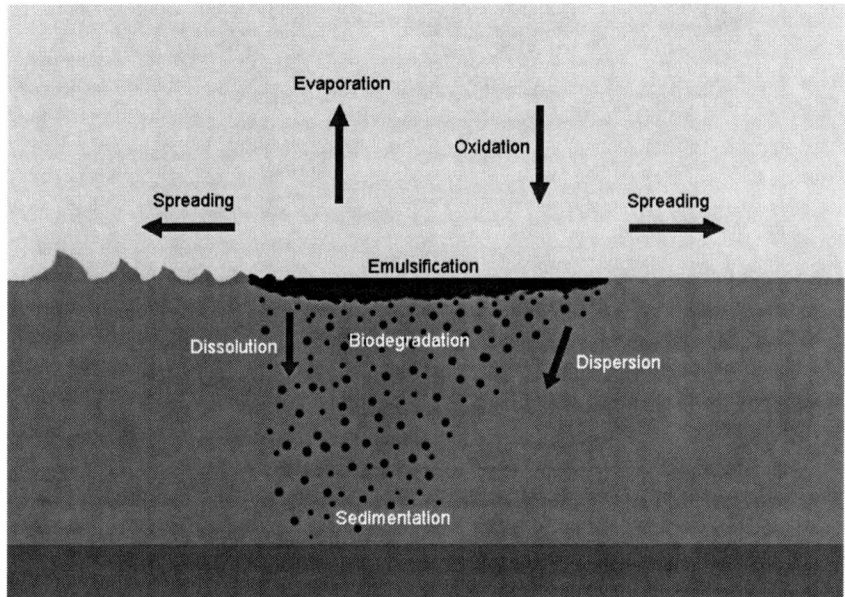

Fig. 2 An illustration of the fate of oil spills in the marine environment

40%. In contrast, heavy or residual oils will lose no more than 10% of their volume in the first days after a spill (National Research Council 2003). The evaporative loss of light hydrocarbons is the main cause of the rapid reduction in volume of the spilled oil, with the subsequent increase in its viscosity and density and the formation of stable oil–water emulsions (Dies et al. 2007). In the case of the *Exxon Valdez*, between 20 and 30% of the spilled oil evaporated in the first few days after the spill; some 40% was washed ashore in Prince William Sound and 7–11% along the coast of the western Gulf of Alaska (Boehm et al. 2007).

Another important process is emulsification: the significance of this process depends on the type of oil and the hydrodynamics (e.g., surface wave turbulence) of the sea waters in question. Emulsification involves the formation of various states of water in oil, often referred to as "chocolate mousse" or "mousse" (National Research Council 2003). It is a process in which water droplets (<0.1 mm in diameter) are incorporated into floating oil. These emulsions may contain 20–80% sea water, thus expanding the volume by 3–5 times the original volume of spilled material (Kingston 2002; National Research Council 2003). The formation of mousse shortly after the spill may be responsible for the long distance transport of less weathered material. Contamination by mousse can also enhance the persistence and retard the weathering of oil (Irvine et al. 2006). For example, in the *Prestige* accident, winds and sea currents drove patches of some 60,000 t of emulsified oil shoreward, affecting more than 800 km of the NW Spanish coast (Gonzalez et al. 2006). Nevertheless, the severe weather conditions, in the days following the spill, carried the emulsified oil back out to sea. These stable emulsions broke up into fragments sufficiently dense that they sank while en route to the shore. In fact, aggregates of tar balls that were 1–20 cm in diameter had densities of up to 300 kg/km^2, and were found in January 2003, when the bottom fauna was sampled by beam trawling in areas of the continental shelf below the main drifting path of the spill (off Costa da Morte; Franco et al. 2006). After the *Amoco Cadiz* disaster, high waves quickly formed a stable water-in-oil emulsion ("mousse") containing 50–70% water. Strong wave activity also rapidly distributed the oil throughout the nearby shore water column. The oil on the surface initially spread eastward as a result of storm winds and tidal currents, until a wind shift 2 weeks after the wreck caused a strong oil movement to the southwest (Gundlach et al.1983).

In the marine environment, solar radiation can cause the photo-oxidation of oil components (Saco-Alvarez et al. 2008). Photo-oxidation is an important process in the weathering of oil that produces a variety of oxidized compounds, including aliphatic and aromatic ketones, aldehydes, carboxylic acids, fatty acids, esters, epoxides, sulfoxides, sulfones, phenols, anhydrides, quinones, and aliphatic and aromatic alcohols. The high polarity and water solubility of these compounds may contribute to the rapid disappearance of an oil slick (Lee 2003). Furthermore, sunlight creates photoproducts that may be more toxic than the original hydrocarbons. For example, photo-oxidized anthracene, benzo(*a*)pyrene, fluoranthene, phenanthrene, and pyrene were found to be more toxic to duckweed (*Lemna gibba*) than to the parent compounds (Lee 2003). The significance of this process, known as phototoxicity, depends on the fuel oil composition, the sensitivity of the exposed organisms and their vital state, the mode of exposure, spectrum, and quality of radiation (Saco-Alvarez et al. 2008).

Dispersion is a mixing process caused by the turbulence field in the ocean (National Research Council 2003). It is the process that would cause a liter of instantaneously released dyed water to expand over time and eventually dissipate in the ocean. Without dispersion, advection would move that liter of dye downstream, but the volume of dyed water would not change over time. Oil spilled on a sea surface can be dispersed by a variety of natural processes; the influence of breaking waves is the dominant process. Breaking waves can split a slick into small droplets, facilitating oil mixing in the water column (Tkalich and Chan 2002). Both horizontal dispersion and vertical dispersion occur, but because the hydrodynamic processes in these two directions are often quite different, a distinction between them is usually made (National Research Council 2003). The vertical dynamics of droplets play a major role in the mass of oil exchanged between the slick and the water column. Wind, waves, and currents speed up the generation of droplets, the smallest of which may be propelled deep into the water column, eventually to be dispersed by the currents (Tkalich and Chan 2002). In addition, the density, viscosity (including the weathering), and thickness of the oil spill, temperature, and salinity of sea water all influence the dispersion processes (Xiankun et al. 1993). Dispersion not only reduces the impact of oil on surface-dwelling animals and enhances biodegradation, but also creates a larger reservoir of oil in the water column, increasing the concentration of dissolved PAHs (polyaromatic hydrocarbons) and the subsequent risks of toxicity to fish (Schein et al. 2009).

The most important process in the self-cleaning of sea waters is biodegradation. This is a microbially mediated process that can result in the partial transformation or complete mineralization of organic compounds and is considered to be one of the major removal mechanisms of organic contaminants (Nam and Kim 2002). An ecosystem's natural capability for self-regeneration, after pollution by oil or oil products, depends not only on the availability of specific oil-oxidizing flora, but also on the presence of other microorganisms and macroorganisms that are able to degrade petroleum hydrocarbons (Tarkhova et al. 2003). Most of the microorganisms present in water are capable of metabolizing a wide range of chemical compounds. Under aerobic conditions, the metabolism of an organic substance involves oxygen, which is used as a hydrogen acceptor. Under conditions in which typical alimentary substrates are absent, bacteria can produce adopted enzymes, as a result of which atypical substrates can be used as the source of carbon for energy generation. Following the biochemical biodegradation of organic substances in the aquatic environment, intermediates of the final products of metabolism are formed.

Almost all bacteria in surface sea water are aerobes, and these bacteria use significant amounts of oxygen for respiration and energy acquisition (Miller and Rutkowska 2002). Among components of petroleum, the C_8–C_{18} alkanes and alkenes are the most easily degraded. The general pattern of aromatic hydrocarbon biodegradation indicates that this process slows with increasing alkylation (Dies et al. 2007). PAHs are resistant to microbiological decomposition, mainly because of the hydrophobic nature of the oil particles, their insolubility in water, and their thermodynamic stability (Klimiuk and Łebkowska 2004). The greater the complexity of the hydrocarbon structure, i.e., the more methyl-branched substitutes or condensed aromatic rings, the slower are the rates of degradation and the greater the likelihood of partially oxidized intermediary metabolites being accumulated. Rates of microbiological degradation vary and are often limited by the concentrations of fixed forms of nitrogen, phosphate, and oxygen in sea waters. Low levels of phosphate and nitrogen are the most frequent factors that serve to limit biodegradation rates in the marine environment (Atlas 1995). The degradation of heavier oil fractions takes a great deal longer and such degradation rates are often assumed to decline exponentially as the hydrocarbons are broken down by light and by microbial consumers. Five years after the *Exxon Valdez* spill, about 2% of the initial oil volume was still lying on the beaches and 13% resided in the sediments, but only a minute fraction was still dispersed in water (Paine et al. 1996).

Under natural marine conditions, the interactions between oil and sediments are important, both as regards the behavior of the contamination and its removal (Delvigne 2002). Bottom sediments are a very significant element of aquatic ecosystems, being ecological niches that support benthic organisms, i.e., the animals and plants living at the bottom of water bodies; the sediment serves as sources of nutrients for aquatic organisms (Wolska and Namieśnik 2002). Oil products are rather severe pollutants because they accumulate in bottom deposits as a result of the high sorption capacity of the sediment-forming particulates. Moreover, they are biochemically highly stable and can accumulate in hydrobionts. The accumulation of oil components in bottom sediments creates conditions favorable to secondary water pollution by oil hydrocarbons and their toxic decomposition products (Belkina

2006). The degree of persistence of oil products in bottom deposits depends on a number of factors, such as the characteristics of the seabed, the temperature, the type of oil product involved, the concentration of nutrients, and the rate of biodegradation (Nikanorov and Stradomskaya 2003). In 1990, over a year after the *Exxon Valdez* oil spill, mean TPAH (total polycyclic aromatic hydrocarbons) concentrations were 4–8 times higher in sediments collected from sites adjacent to heavily oiled shorelines than at reference sites (Table 3; Jewett et al. 1999).

Table 3 Mean concentrations of TPAH (total polycyclic aromatic hydrocarbons) in shallow subtidal sediments at 6–20 m and below 3 m, in oiled (O) and reference (R) sites of Prince William sound

Year	BI (O)	DB (R)	HB (O)	LHB (R)	SB (O)	MB (O)	Oiled	Reference
TPAH (ng/g); 6–20 m								
1990	15,253	1,837	45	1,908	872	102	5,390	1,282
1991	1,122	130	121	103	305	116	116	116
1993	612	95	87	0	117	74	272	57
1995	256	66	34	118	377	121	222	101
TPAH (ng/g); <3 m								
1990	14,309	1,418	322	297	306	124	4,979	613
1991	790	915	137	60	174	59	367	345
1993	443	59	40	45	11	42	164	49
1995	313	179	20	9	16	60	116	83

BI: Bay of Isles, DB: Drier Bay, HB: Herring Bay, LHB: Lower Herring Bay, SB: Sleepy Bay, MB: Moose Lips Bay (Jewett et al. 1999)

4 The Impact of Oil Spills on Seabirds, Fish, and Marine Animals

Oil spills have both acute (rapid, short-term) and chronic (long-lasting) effects on ecosystems. The flora and fauna most at risk are those coming into direct contact with recently spilled oil. Many of the chemicals in oil spills are toxic and can thus have devastating effects on plankton, fish, and animals living on the seabed.

The impact of an oil spill on biota depends on a number of factors, such as:

– the rate of spread of the oil slick;
– the oil composition;
– the location of the spill;
– the time or season of the accident (bird migrations);
– the properties, toxicity, and stability of the petroleum substance;
– the species biodiversity at the site of the oil spill;
– environmental sensitivity, i.e., proximity of bird habitat, beaches, rocks, wetlands; and
– the number and type of habitats.

Three major pathways lead to long-term impacts:

- the chronic persistence of oil, biological exposure, and population affect the species that are closely associated with shallow sediments;
- the delayed population effects of sublethal doses that may compromise health, growth, and reproduction; and
- the indirect effects of trophic interaction cascade, all of which transmit effects well beyond the acute-mortality phase (Peterson et al. 2003).

One of the most obvious mineral oil-related adverse effects of (chronic) pollution of the world's oceans and seas is the mortality of seabirds (Camphuysen and Heubeck 2001). Seabirds are highly vulnerable to "surface pollutants," but in fact this term embraces a wide range of lipophilic substances, including mineral oils, vegetable oils, and various other chemicals. Mineral oils pose the greatest threat to seabirds, by far, but several incidents have demonstrated the equally lethal effects of many other substances (Camphuysen and Heubeck 2001). The primary effect on birds of oil contamination is the loss of body insulation that is provided by the feathers: the cold water reaches the skin, leading to hypothermia and death. Furthermore, large amounts of oil cause the feathers to stick together, impairing flight and buoyancy. Birds may ingest and/or inhale oil while trying to preen or eat contaminated food. Consequently, they suffer rapid, short-term or long-term effects, such as damage to the lungs, kidneys and liver, and gastro-intestinal disorders. For example, between 80,000 and 150,000 seabirds that over winter in the Bay of Biscay were killed during the *Erika* oil spill (Cadiou et al. 2004). Unprecedented numbers of seabird deaths (estimated at 250,000) were documented during the days after the *Exxon Valdez* oil spill (Peterson et al. 2003). More than 4,000 birds died in the immediate aftermath of the *Prestige* oil spill and more than 40,000 affected birds were subsequently collected. Oil may also be transferred from the birds' plumage to the eggs they brood, smothering the eggs by sealing the pores in the shells and preventing the exchange of gases (EPA 1999).

After oil has entered the marine environment, it forms a multimillimeter thick slick covering the sea surface as a microlayer (Guitart et al. 2008). The slick hinders gas exchange with the air and limits penetration of solar radiation (Page et al. 2000). This substantially slows down the rate of photosynthesis, thereby reducing the populations of many marine plants and sea organisms. For example, after the *Exxon Valdez* spill, flowering and shoot density were reduced in eelgrass *Zostera marina* population, although they managed to recover within 2 years of the spill. There were catastrophic declines in benthic fauna from anoxia in a heavily oiled fjord several months after an oil spill (Jewett et al. 1999).

Fish may be exposed to spilled oil in different ways. The water column may contain toxic and volatile components of oil that may be absorbed by their eggs and juvenile stages; and they may consume contaminated food. Direct contact with oil causes blockage of the gills. Fish exposed to oil may suffer from changes in heart and respiratory rates, enlarged livers, reduced growth, fin erosion, as well as a variety of biochemical and cellular changes, and reproductive and behavioral responses

(EPA 1999). Therefore, the consequence of such exposure may be disease or death. In addition, toxic substances are bioaccumulated in the tissues of marine organisms and then biomagnified up the marine food chain from phytoplankton to fish to seals and other sea mammals. In this way, birds and other predators at the top of the food pyramid are exposed to the substances that were absorbed at the lower levels of the pyramid. It is estimated that conversion at a higher trophic level magnifies the quantity of accumulated toxic substances by a factor of 3–5. Toxic substances are transferred not only from one species to another, but also from one generation to another of the same species.

The accumulation of petroleum hydrocarbons by marine organisms depends on the degree of bioavailability of hydrocarbons, the length of exposure, and the organism's capacity for metabolic transformations (National Research Council 2003). This accumulation process is enhanced because hydrocarbons and other components of oil are primarily lipophilic and are easily assimilated by organisms. Mussels, for example, are exposed as a result of their filter-feeding mechanisms: mussels have been used as an indicator species for the presence of petroleum hydrocarbons, because they have a greater potential than other marine organisms to accumulate petroleum components in their tissues. Such contaminants become concentrated in mussels at levels higher than ambient levels, but when the animals are returned to clean water, depuration is equally rapid, with tissue levels falling to less than 10% of the peak value within 8 d, and almost back to background levels in 16 d (Kingston 2002).

Like seabirds, mammals may also come into contact with oil on the sea surface. The animals affected include river otters, beavers, sea otters, polar bears, manatees, seals, sea lions, walruses, whales, porpoises, and dolphins. The sensitivity of mammals to spilled oil is highly variable (EPA 1999). The amount of damage appears to be most directly related to how important the fur and blubber are to thermoregulation; therefore, hypothermia is a serious problem in the case of an oil spill. Because oil contamination imparts to fish and other animals unpleasant tastes and smells, predators will sometimes refuse to eat their prey and will begin to starve. Sometimes a local population of prey organisms is destroyed, because they have no food resources after an oil spill (EPA 1999).

5 Shipping Accidents in the Baltic Sea

Shipping traffic in the Baltic Sea is quite heavy – 15% of the world's cargo moves on it. It has been estimated that at any moment there are about 2,000 ships transporting their cargos or passengers in the Baltic, including some 200 large tankers and other vessels transporting hazardous substances. Hence, any of these vessels have the potential to cause environmental contamination. In 2006, 60% of these ships were cargo vessels, 18% were tankers, and 11% were passenger ships (HELCOM BSEP (Baltic Sea Environment Proceedings) 2008). In 2000, 80 million t of crude oil and petroleum products were transported across the Baltic (HELCOM BSEP

2003). By 2007, this amount had risen to >170 million t and is still increasing (HELCOM BSEP 2008).

The Baltic Sea has not suffered from oil spills as catastrophic as the accidents of the *Prestige* or *Amoco Cadiz*. Nevertheless, oil remains a serious threat to Baltic ecosystems and wildlife (HELCOM BSEP 2003). The location of the Baltic, the low intensity of its self-cleaning processes, the relatively shallow depth, and the slow rate of exchange of water with the North Sea (and hence the world oceans) mean that even a small leakage could be disastrous for the environment. Oil decomposes slowly in the cold waters of the Baltic, where the average water temperature is only about 10°C (HELCOM BSEP 2003). Moreover, there are navigational hazards from the large number of islands that populate the Baltic, and the long periods during the year when Baltic waters are frozen over. The environmental impact of shipping in the Baltic is becoming more marked as a result of the intensification of sea traffic there and also the increasing size of the vessels using this sea.

There is one additional and important point that speaks to the sensitivity of this body of water. Compared to other aquatic ecosystems, relatively few animal and plant species live in the brackish ecosystems of the Baltic Sea; even with this limited diversity there is a unique mix of marine and freshwater species that are adapted to the Baltic's brackish conditions, as well as a few true brackish water species. The limited number of species involved in Baltic-Sea-food webs means that each individual species has a special importance in terms of their connection to the structure and dynamics of the whole ecosystem. The disappearance of a single key species could destroy the whole system (HELCOM BSEP 2003).

Between 1969 and 1995 there were 40 incidents in Baltic waters involving leakages from vessels of at least 100 t of oil and associated products (HELCOM 2006). Table 4 presents a list of these worst oil-spill accidents (HELCOM 2006). The most common types of accidents were groundings, collisions, technical failures, and fires/explosions. Table 5 sets out the most common types of accidents that have occurred in the Baltic Sea (HELCOM BSEP 2008; HELCOM 2008).

Table 4 The worst shipping accidents in the Baltic Sea that have resulted in oil spills (HELCOM 2006)

Year	Name of ship	Location	Oil lost (t)
1969	*Benedicte*	Trelleborg, Sweden	2,700
1970	*Irini*	Nynashamn, Sweden	1,000
1973	*Jawachla*	Trelleborg, Sweden	1,500–2,000
1977	*Tsesin*	Nynashamn, Sweden	1,000
1979	*Antonio Gramsci*	Ventspils, Latvia	5,500
1981	*Globe Asimi*	Klaipeda, Lithuania	16,000
1981	*Jose Martin*	Dalaro, Sweden	1,000
1990	*Volgoneft*	Karlskrona, Sweden	1,000
1995	*Hual Trooper*	The Sound, Sweden	180
2001	*Baltic Carrier*	Kadetrenden, Denmark	2,700
2003	*Fu Shan Hai*	Bornholm, Denmark/Sweden	1,200

Table 5 Number and types of reported shipping accidents in the Baltic Sea, 2000–2007 (HELCOM BSEP 2008; HELCOM 2008)

Types of accidents	2000	2001	2002	2003	2004	2005	2006	2007
Grounding	36	37	41	30	57	53	46	54
Collision	11	11	9	24	44	54	54	40
Other	15	8	11	17	41	39	17	26
Total	62	56	61	71	142	146	117	120

Numerous accidents have taken place in the Danish Straits, the Gulf of Finland, and in harbors. Fortunately, few of the accidents in the Baltic have led to catastrophic pollution, but even one large-scale accident would gravely threaten the marine environment (HELCOM BSEP 2008). The accidents with the most far-reaching effects in the last 10 years were those involving the *Baltic Carrier* and the *Fu Shan Hai*. The oil tanker *Baltic Carrier* was sailing from Muga, Estonia, carrying a cargo of heavy fuel oil, when it collided with the bulk carrier *Tern,* east of the Danish island of Falster, on March 29, 2001. As a result, the *Baltic Carrier* sprang a leak, and ca. 2,700 t of the total cargo of 33,000 t of heavy fuel oil flowed into the sea. The oil, which had the viscosity of floating asphalt, drifted toward the islands of Møn and Falster (HELCOM SEA 2002). Despite logistical and organizational problems, not to mention bad weather, the clean-up effort by the Danish authorities was successful. Unfortunately, some 2,000 birds were killed, and from 4,000 to 4,500 will have been affected by the oil in one way or another (HELCOM SEA 2002). The Danish authorities also carried out the rescue operation after the collision between the bulk carrier *Fu Shan Hai* and the container ship *Gdynia*, which occurred on May 31, 2003, off the island of Bornholm. As a result, >1,200 t of oil products leaked out of the *Fu Shan Hai*. The very competently coordinated recovery operation on the part of the Danish, Swedish, and German oil-slick response teams was able to limit the damage to the environment (HELCOM RESPONSE 2003). Even so, the oil pollution persisted for a month, and 1,100–1,600 seabirds, mainly auks and eiders, were killed (HELCOM NEWS 2006).

Illegal discharges of oil from ships have been detected in the Baltic. The Baltic Area – protected by the International Convention for the Prevention of Pollution from Ships (MARPOL (International Convention for the Prevention of Pollution from) 73/78) – is a "Special Area" (Annex I – Oil, Annex V Garbage, Annex VI – Prevention of air pollution by ships). MARPOL defines certain sea areas as "Special Areas" in which, for technical reasons relating to their oceanographic and ecological condition and to their sea traffic, the adoption of special mandatory methods for the prevention of marine pollution is required (IMO (International Maritime Organization) 2009). As a result, the discharge into the Baltic Sea of oil and diluted mixtures containing oil in any form, including crude oil, fuel oil, oil sludge, or refined products, is forbidden. Despite this prohibition, however, 238 illicit oil spills were detected during a total of 3,969 hr of surveillance flights conducted by Baltic coastal countries over the Baltic Sea during 2007, compared to 236 discharges

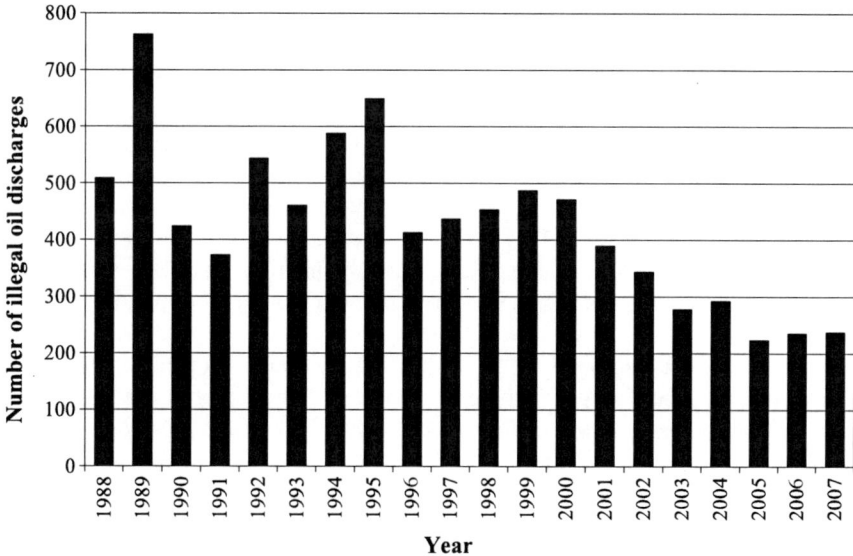

Fig. 3 The annual number of detected illegal oil discharges in the Baltic Sea, 1988–2007 (HELCOM 2006, HELCOM BSEP 2008)

observed during 5,128 air patrol hours, in 2006. Figure 3 presents information on the number of detected illegal oil discharges in the Baltic Sea during the period 1988–2007 (HELCOM 2006, HELCOM BSEP 2008).

To keep things in perspective, it should be borne in mind that the number of illegal oil discharges has been decreasing steadily since the 1980s. In view of the unique and rare Baltic ecosystems and their vulnerability to degradation by human activities, the HELCOM member countries, in support of a 2003 Swedish initiative, submitted a proposal to the IMO that Particularly Sensitive Sea Area (PSSA) status be conferred on the Baltic Sea. A PSSA is an area that needs special protection through action by the IMO because of its significance for recognized ecological or socio-economic or scientific reasons and which may be vulnerable to damage by international maritime activities. The IMO approved this proposal in 2004, which resulted in the Baltic Sea being placed among the most valuable and sensitive marine ecosystems on the world, alongside the Great Barrier Reef, Canary Islands, and the Galapagos Archipelago.

6 Implications of Shipping-oil Spills

Tanker accidents have drawn the attention of the international community to the problem of oil pollution, its complexity, and consequences. They have also demonstrated the existence of risk and the absence of regulations and procedures stipulating the action to be taken in the event of an oil spill. This has led to the creation of an international legal framework that covers ship construction, navigational safety, crew training, and responsibility for environmental damage. When analyzing the reasons for formulating these international laws and maritime strategies, one can see that the international community becomes aware of the hazards of transporting oil by sea only after a catastrophic event causes massive environmental damage. To wit, the accident of the *Torrey Canyon* led to the adoption of the International Convention Relating to Intervention on the High Seas in Cases of Oil Pollution Casualties (INTERVENTION), the International Convention on Civil Liability for Oil Pollution Damage (CLC), and the International Convention on the Establishment of an International Fund for Compensation for Oil Pollution Damage (FUND). One of the most visible effects of the *Exxon Valdez* incident was the enactment of the Oil Pollution Act (OPA) of 1990. After the accident of the tanker *Erika,* in 1999, the European Commission stated that an appropriate law should be *"designed to bring about a change in the prevailing mentality in the seaborne oil trade. More powerful incentives are needed in order to persuade the carriers, charterers, classification societies and other key bodies to give a higher profile to quality considerations. At the same time, the net should be tightened on those who pursue short-term personal financial gain at the expense of safety and the marine environment."* (Europa Summaries of EU Legislation 2007). The effect was the drawing up of the three so-called *Erika* packages, the aims of which included the improvement of safety at sea to be achieved by the setting up of a system of identifying and monitoring ships (Directive 2002/59/EC), monitoring the activity of classification societies (Directive 2001/105/EC), inspection of foreign ships entering European Union ports (so-called Port State Control), the withdrawal of single-hull tankers (Regulation (EC) 417/2002), and the establishment of a European Maritime Safety Agency.

These examples provide ample demonstration that new, more rigorous standards, aimed at minimizing the probability of accidents, are enacted only in the face of real danger. One has to remember, however, that the introduction of stricter regulations is often very difficult for reasons of conflicting interests. International conventions are thus frequently a compromise between national interests and the need to protect the environment on a global scale. Undoubtedly, the improvement of the sea and ocean water quality is connected first to the incremental ecological awareness of society and cooperation at the international and regional levels. Although military-related cooperation between countries with different political ideologies is not usually possible, the mutual benefits of ecological cooperation between countries with different political regimes are possible. For example, at the Convention on the Protection of the Marine Environment of the Baltic Sea Area 1992 (The Helsinki Convention 1992) a cooperative agreement was signed by the then seven Baltic coastal states. This agreement was a huge success despite the fact that the participating countries belonged to blocks of opposite ideological, political, and economic postures. Moreover, for the first time ever, all sources of pollution of the entire sea were made subject to the outcome of a single convention.

7 Summary

Since ancient times, ships have sunk during storms, either as a result of collisions with other vessels or running onto rocks. However, the ever-increasing importance of crude oil in the twentieth century and the corresponding growth in the world's tanker fleet have drawn attention to the negative implications of sea transport. Disasters involving tankers like the *Torrey Canyon* or the *Amoco Cadiz* have shown how dramatic the consequences of such an accident may be.

The effects of oil spills at sea depend on numerous factors, such as the physicochemical parameters of the oil, the characteristics of the environment affected, and the physical, chemical, and biological processes occurring there, such as evaporation, dissolution, dispersion, emulsification, photo-oxidation, biodegradation, and sedimentation. The combination of these processes reduces the concentrations of hydrocarbons in sediments and water and alters the chemical composition of spilled oils. In every case, oil spills pose a danger to fauna and flora and cause damage to sea and shores ecosystems. Many of the petroleum-related chemicals that are spilled are toxic, otherwise carcinogenic or can be bioaccumulated in the tissues of marine organisms. Such chemicals may then be biomagnified up the marine food chain from phytoplankton to fish, then to seals and other carnivorous sea mammals. Moreover, oil products can be accumulated and immobilized in bottom deposits for long periods of time. Oil spills are particularly dangerous when they occur in small inland seas that have intense sea traffic, e.g., the Baltic Sea.

References

Albaigés J, Morales-Nin B, Vilas F (2006) The Prestige oil spill: A scientific response. Editorial Mar Pollut Bull 53:205–207

Atlas RM (1995) Petroleum biodegradation and oil spill bioremediation. Mar Pollut Bul 4–12(31):178–182

Bądkowski A (1977) Problemy zanieczyszczenia morza cz. 1. Wydawnictwo Instytutu Morskiego w Gdańsku, Gdańsk – Szczecin, pp 5–10

Belkina NA (2006) Pollution of bottom sediments in Petrozavodsk Bay of Lake Onega with oil products. Water Resour 2(33):163–169

Boehm PD, Neff JM, Page DS (2007) Assessment of polycyclic aromatic hydrocarbon exposure in the waters of Prince William Sound after the *Exxon Valdez* oil spill: 1989–2005. Mar Pollut Bull 54:339–367

Borakowski H (1955) Zanieczyszczenie morza produktami naftowymi. Technika i Gospodarka Morska 9:234–236

Cadiou B, Riffaut L, McCoy KD, Cabelguen J, Fortin M, Gelinaud G, Le Roch A, Tirard C, Boulinier T (2004) Ecological impact of the "Erika" oil spill: Determination of the geographic origin of the affected common guillemots. Aquat Living Resour 17:369–377

Camphuysen CJ, Heubeck M (2001) Marine oil pollution and beached bird surveys: the development of a sensitive monitoring instrument. Environ Pollut 112:443–461

Camphuysen CJ (2007) Chronic oil pollution in Europe. IFAW Report, Royal Netherlands Institute for Sea Research. IFAW Publication, UK

Delvigne GAL (2002) Physical appearance of oil in oil-contaminated sediment. Spill Sci Tech Bull 1(8):55–63

Dies S, Jover E, Bayona JM, Albaiges A (2007) *Prestige* oil spill. III. Fate of a heavy oil in the marine environment. Environ Sci Technol 41:3075–3082

E.I.A. (2009) Energy Information Administration Official Energy Statistics from the U.S. Government.http://tonto.eia.doe.gov/cfapps/ipdbproject/iedindex3.cfm?tid=5&pid=54&aid=2&cid=&syid=1980&eyid=2007&unit=TBPD

EMSA (2009) The world merchant fleet in 2007. Statistics from Equasis. EMSA (The European Maritime Safety Agency) Publication. https://extranet.emsa.europa.eu/index.php?option=com_docman&task=cat_view&gid=133&Itemid=193

EPA (1999) Understanding oil spills and oil spill response. Emergency response. EPA report 540-K-99-007

Europa Summaries of EU Legislation (2007) http://europa.eu/legislation_summaries/transport/waterborne_transport/l24230_en.htm

Franco MA, Vinas L, Soriano JA, De Armas D, Gonzalez JJ, Beiras R, Salas N, Bayona JM, Albaiges J (2006) Spatial distribution and ecotoxicity of petroleum hydrocarbons in sediments from the Galicia continental shelf (NW Spain) after the Prestige oil spill. Mar Pollut Bull 53:260–271

Gelin A, Gravez V, Graham JE (2003) Assessment of *Jessica* oil spill impacts on intertidal invertebrate communities. Mar Pollut Bull 46:1377–1384

GESAMP (2007) Estimates of oil entering the marine environment from sea-based activities. Report and Studies GESAMP No. 75, GESAMP publication, London

Gonzalez JJ, Vinas L, Franco MA, Fumega J, Soriano JA, Grueiro G, Muniategui S, Lopez-Mahia P, Prada D, Bayona JM, Alzaga R, Albaiges J (2006) Spatial and temporal distribution of dissolved/dispersed aromatic hydrocarbons in seawater in the area affected by the *Prestige* oil spill. Mar Pollut Bull 53:250–259

Gundlach ER, Boehm PD, Marchand M, Atlas RM, Ward DM, Wolfe DA (1983) The fate of *Amoco Cadiz* oil. Science 221:122–229

Guitart C, Frickers P, Horrillo-Caraballo J, Law RJ, Readman JW (2008) Characterization of sea surface chemical contamination after shipping accidents. Environ Sci Technol 7(42): 2275–2282

HELCOM (2006) Maritime transport in the Baltic sea. Draft HELCOM Thematic Assessment in 2006, HELCOM, Finland

HELCOM (2008) Report on shipping accidents in the Baltic Sea area for the year 2007. Baltic Marine Environment Protection Commission, HELCOM, Finland

HELCOM BSEP (2003) The Baltic Marine Environment 1999–2002. Baltic Sea environment Proceedings No. 87, HELCOM, Finland

HELCOM BSEP (2008) Activities 2007 overview. Baltic Sea Environment Proceedings No. 114, HELCOM, Finland

HELCOM NEWS (2006) http://www.helcom.fi/press_office/news_baltic/en_GB/1143707191730/?u4.highlight=Fu%20shan%20hai

HELCOM RESPONSE (2003) Oil and other harmful substances: The Collision between Chinese Bulk Carrier Fu Shan Hai and Cypriot Container Vessel Gdynia on 31 May 2003. HELCOM RESPONSE No 3, HELCOM, Finland

HELCOM SEA (2002) Baltic carrier incident and the response to the spill. HELCOM, Finland

IMO (2009) www.imo.org

Irvine GV, Mann DH, Short JW (2006) Persistence of 10-year old *Exxon Valdez* oil on Gulf of Alaska beaches: The importance of boulder-armoring. Mar Pollut Bull 52:1011–1022

ITOPF (2002) Fate of oil spill. ITOPF Technical Information Paper, ITOPF, London

ITOPF (2009) ITOPF handbook 2009/2010. ITOPF, London

Jewett SC, Dean TA, Smith RO, Blanchard A (1999) '*Exxon Valdez*' oil spill: Impacts and recovery in the soft-bottom benthic community in and adjacent to eelgrass beds. Mar Ecol Progr 185: 59–83

Kachel JM (2008) particularly sensitive sea areas. The IMO's role in protecting vulnerable marine areas: Threats to the marine environment: Pollution and physical damage, Springer, Berlin, Heidelberg, pp 13–26

Klimiuk E, Łebkowska M (2004) Biotechnologia w ochronie środowiska. Wydawnictwo Naukowe PWN, Warszawa, pp 204–213

Kingston PF (2002) Long-term environmental impact of oil spills. Spill Sci Tech Bull 1–2(7): 53–61

Kingston PF, Runciman D, McDougall J (2003) Oil contamination of sedimentary shores of the Galapagos Islands following the wreck of the *Jessica*. Mar Pollut Bull 47:303–312

Lee RF (2003) Photo-oxidation and photo-toxicity of crude and refined oils. Spill Sci Tech Bull 2(8):157–162

Łopuski J (1974) Prawo morskie dla oficerów marynarki wojennej i rybołówstwa. Wydawnictwo Morskie, Gdańsk, 188 pp

Maki WA (1991) The *Exxon Valdez* oil spill: Initial environmental impact assessment. Environ Sci Technol 1(25):24–29

Meadows DH, Meadows DL, Randers J, Behrens WW (1973) The limits to growth. Państwowe Wydawnictwo Ekonomiczne, Warszawa [in polish].

Miller H, Rutkowska M (2002) Oil pollution, a new concept of biodegradability determination of noxious liquid substances transporter by the Polish sea area. In: Hupka J (ed) Oil pollution: Prevention, characterization, clean technology: Proceedings of 3rd international conference held at the Gdansk University of technology, Gdansk, Poland, September 8–11, Gdańsk, pp 41–45

Nam K, Kim JY (2002) Persistence and bioavailability of hydrophobic organic compounds in the environment. Geosci J 1(6):13–21

National Research Council (1985) Steering Committee for the Petroleum in the Marine Environment Update, Board on Ocean Science and Police, Ocean Sciences Board, Commission on Physical Sciences, Mathematics, and Resources. Oil in the Sea: Inputs, Fates, and Effects. National Academy Press, Washington, DC, pp 273–274

National Research Council (2003) Committee on Oil in the Sea: Inputs, Fates, and Effects, Ocean Studies Board and Marine Board, Divisions of Earth and Life Studies and Transportation Research Board. Oil in the Sea III: Inputs, Fates, and Effects. National Academy Press, Washington, DC, pp 90–103

Nikanorov AM, Stradomskaya AG (2003) Oil products in bottom sediments of freshwater bodies. Water Resour 1(30):98–102

Page CA, Bonner JS, Summer PL, Autenrieth RL (2000) Solubility of petroleum hydrocarbons in oil/water systems. Mar Chem 70:79–87

Paine RT, Ruesink JL, Sun A, Soulanille EL, Wonham MJ, Harley ChDG, Brumbaugh DR, Secord DL (1996) Trouble on oiled waters: Lessons from the *Exxon Valdez* oil spill. Ann Rev Ecol Syst 27:197–235

Peterson ChH, Rice SD, Short JW, Esler D, Bodkin JL, Ballachey BE, Irons DB (2003) Long-term ecosystem response to the *Exxon Valdez* oil spill. Science 302:2082–2086

Pietraszek R (1967) Problemy zanieczyszczenia mórz olejami wynikłe z katastrofy zbiornikowca, *Torrey Canyon*. Technika i Gospodarka Morska 8–9:373–376

Różańska Z (1987) Zasoby, zanieczyszczenia i ochrona wód morskich ze szczególnym uwzględnieniem Bałtyku. Państwowe Wydawnictwo Naukowe, Warszawa, pp 83–84

Saco-Alvarez L, Bellas J, Nieto O, Bayona JM, Albaiges J, Beiras R (2008) Toxicity and phototoxicity of water-accommodated fraction obtained from *Prestige* fuel oil and marine fuel oil evaluated by marine bioassays. Sci Total Environ 394:275–282

Surygała J (2001) Ropa naftowa, a środowisko przyrodnicze. Oficyna Wydawnicza Politechniki Wrocławskiej, Wrocław, pp 32–39

Schein A, Scott JA, Mos L, Hodson PV (2009) Oil dispersion increases the apparent bioavailability and toxicity of diesel to Rainbow Trout (*Oncorhynchus mykiss*). Environ Toxicol Chem 28(3):595–602

Tarkhova E, Koval'Chuk YL, Poltarukha OP (2003) A study of oil biodegradation in the black sea water. Water Resour 1(30):92–97

The Helsinki Convention 1992. The Convention on the Protection of the Marine Environment of the Baltic Sea Area, 1992. www.helcom.fi

Tkalich P, Chan ES (2002) Vertical mixing of oil droplets by breaking waves. Mar Pollut Bull 44(11):1219–1229

Wang Z, Fingas M, Page DS (1999) Oil spill identification. J Chromatogr A 843:369–411

Wiewióra A, Wesołek Z, Puchalski J (2007) Ropa naftowa w transporcie morskim. TRADEMAR, Gdynia, pp 6–10

Wolska L, Namieśnik J (2002) Distribution of Pollutants in the Odra River System Pol. J Environ Stud 6(11):663–668

Xiankun L, Jing L, Shuzhu Ch (1993) Dynamic model for oil slick dispersion into a water column – A wind-driven wave tank experiment. Chin J Oceanol Limnol 2(11):161–170

Water Quality in South San Francisco Bay, California: Current Condition and Potential Issues for the South Bay Salt Pond Restoration Project

J. Letitia Grenier and Jay A. Davis

Contents

1	Introduction	115
	1.1 San Francisco Estuary	116
	1.2 Salt Ponds and Wetland Restoration	117
2	Current Water Quality	120
	2.1 Chemical Contaminants	120
	2.2 Other Aspects of Water Quality	130
3	Potential Future Changes Related to Restoration	131
	3.1 Erosion of Contaminants at Depth	131
	3.2 Change of Habitat Types	134
4	Future Changes that May Affect Water Quality	138
5	Recommendations	139
6	Summary	141
	References	142

1 Introduction

Reengineering of the natural world is a hallmark of the human species. Along with this reengineering comes a need to sometimes reverse previous modifications. Management of wetlands in the USA is one example of this cycle of modification and restoration. Loss of wetlands across the USA during European colonization and industrialization has been followed decades and even centuries later by efforts to restore many of these habitats. Habitats can never be restored to their original, pristine form and function, and complete restoration is even more difficult in highly modified landscapes that have large human populations. In this chapter, we address

J.L. Grenier (✉)
San Francisco Estuary Institute, 7770 Pardee Lane, Oakland, CA 94621, USA
e-mail: letitia@sfei.org

water quality concerns that derive from chemical contamination that is related to the restoration of a large area of former tidal wetlands in the highly urbanized San Francisco Estuary in California, USA. Other water quality concerns are also briefly addressed. We begin by describing the San Francisco Estuary and cogent background associated with the South Bay Salt Pond Restoration Project.

1.1 San Francisco Estuary

An unprecedented extent of tidal wetland restoration is ongoing and planned in San Francisco Bay and the greater Estuary, henceforth referred to as the "Estuary" (www.californiawetlands.net/tracker/). The Estuary is a series of large embayments fed by the Sacramento and San Joaquin rivers, which drain California's Central Valley and the surrounding Sierra Nevada mountains. Smaller, local watersheds also contribute to the waters that exit the Estuary through the Golden Gate at the mouth of San Francisco Bay (Fig. 1). In total, the Golden Gate watershed comprises nearly 40% of the area of California and includes a vast variety of land uses. Principal among these uses are agriculture, urban and industrial areas, and cattle ranching (Goals Project 1999). Freshwater inputs to the Estuary from the major rivers are reduced by diversions for agriculture and urban use. These diversions vary in wet and dry years from 10% to nearly 75% of the volume of water that flows into the Bay (URS 2007).

Water quality in the Estuary is affected not only by how much water is diverted from entering it, but also by the quality of the waters that do enter the Estuary from the surrounding watershed. The quality of the incoming waters depends on land use and water quality management in the surrounding watersheds. The local watershed of the Estuary is home to the cities of San Francisco, San Jose, and Oakland as well as high-tech industry in Silicon Valley. Therefore, the local watershed is characterized by urban development, especially toward the south. In 2005, more than 7 million people resided in the Bay Area, and by 2035, a population of 9 million is expected (ABAG 2007). Although the Bay Area has produced advances in many areas of human endeavor, including technology, social values, and environmental consciousness, it has also produced a legacy of contamination in its waterways. Mercury, polychlorinated biphenyls (PCBs), and a host of other chemical contaminants are prevalent in the Bay and in the surrounding wetlands. Many of these contaminants have local urban sources, and others have distant sources, such as mines in the Sierra Nevada or coal-fired power plants in China.

Although the human population of the Bay Area has become urbanized and less directly dependent on the local landscape in recent decades, abundant biological resources and ecological services were important attractions for the original settlement and industrialization of the region. Early accounts of the Estuary describe an ecological jewel characterized by thriving fisheries, oak woodlands teeming with wildlife, and vast marshes stretching from Bay to upland (Goals Project 1999). The Estuary has a plethora of biological diversity and retains many wildlife taxa found

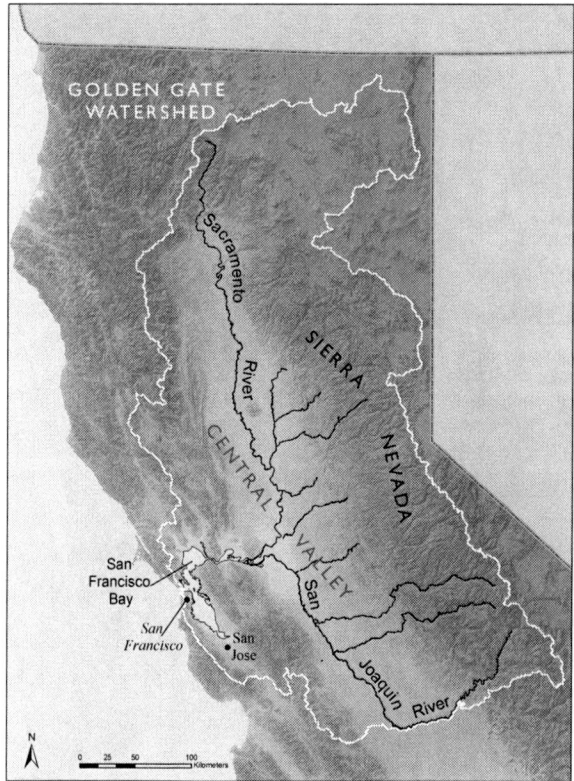

Fig. 1 San Francisco Bay and its watershed, which comprises much of northern California, USA. The Bay is connected to the Pacific Ocean at the Golden Gate, a narrow opening about mid-way down the western side of the Bay (see Fig. 5 for location). The San Francisco Estuary includes an inland delta ("the Delta") where the Sacramento and San Joaquin rivers meet. The maps in Fig. 5 show the approximate spatial extent of the Estuary

nowhere else. However, several species that were once abundant in the Estuary have declined to the point that they now require protected status from state and federal agencies. The tidal wetlands, in particular, are home to several special-status, endemic taxa, such as the California Clapper Rail (*Rallus longirostris obsoletus*) and saltmarsh harvest mouse (*Reithrodontomys raviventris raviventris*), both of which are endangered species (Federal Register 1970).

1.2 Salt Ponds and Wetland Restoration

More than 85% of the tidal wetland acreage (fresh, brackish, and salt marsh) of the San Francisco Estuary has been lost to human alteration of the landscape since Europeans settled the area (Fig. 2) (Goals Project 1999). Many Estuary marshes

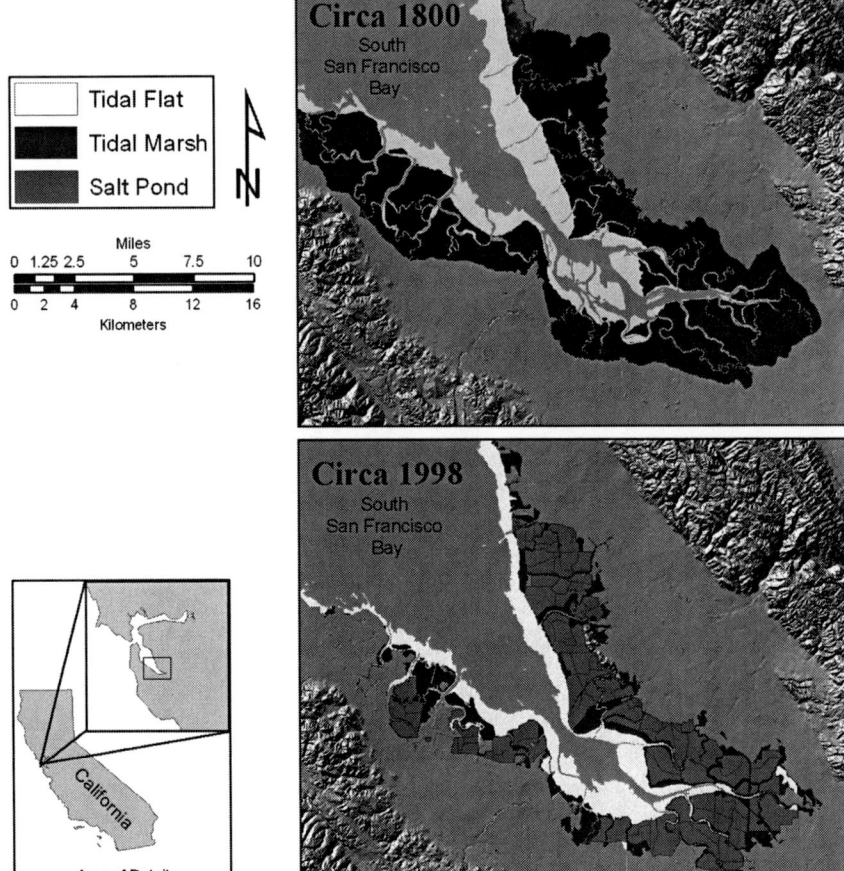

Fig. 2 Tidal marsh habitats in South San Francisco Bay circa 1998 were reduced greatly in acreage, compared to their extent in 1800. In this region of the San Francisco Estuary, most of the former marshes have been converted to salt ponds (Goals Project 1999)

were diked for agriculture, ranching, or urban and industrial development. Other marshes, in San Pablo Bay and South San Francisco Bay, in particular, were converted to salt evaporation ponds. Nearly the entire perimeter of South Bay, which was historically fringed by marshes several miles wide in some areas, was converted to salt ponds (Fig. 2). These salt ponds were developed gradually over decades, from 1857 to 1960 (Collins and Grossinger 2004). The ponds were managed to produce industrial-grade salt by trapping estuarine water in the ponds nearest the Bay and gradually concentrating the salts in the water through evaporation. Ultimately, the highly saline water would be completely evaporated in the ponds closest to land, and then the salt was harvested.

An effort to protect the Bay began in the 1960s, when inadequate treatment of sewage led to regular fish die-offs. Recognition of the importance of restoring local estuarine ecosystems followed, and, in the 1990s, culminated in a consortium of government agencies, non-governmental organizations, scientists, and private citizens working together on the Bay Area Wetlands Ecosystem Goals Project. The purpose of the "Goals Project" was to identify the quantity, type, and distribution of wetlands needed to sustain diverse communities of estuarine wildlife; the goals also included the task of performing an analysis of the pre-industrialization extent and function of Bay Area tidal wetlands (Goals Project 1999).

Wetlands restoration boomed following the publication of the Bay Area Wetlands Ecosystem Goals (Goals Project 1999); the largest single effort is the South Bay Salt Pond Restoration Project (SBSPRP). This vast project comprises 15,100 A in South San Francisco Bay (Fig. 3). The SBSPRP aims to restore and enhance a mix of wetland habitats, provide for flood management, and enhance public access and recreation opportunities (www.southbayrestoration.org). The Project came into being following public acquisition of the ponds from the Cargill Company and transfer of ownership and management to the US Fish and Wildlife Service and the California Department of Fish and Game.

The first stages of the SBSPRP are underway. The pond purchase from Cargill was finalized in 2003, and an initial stewardship plan has been in operation since

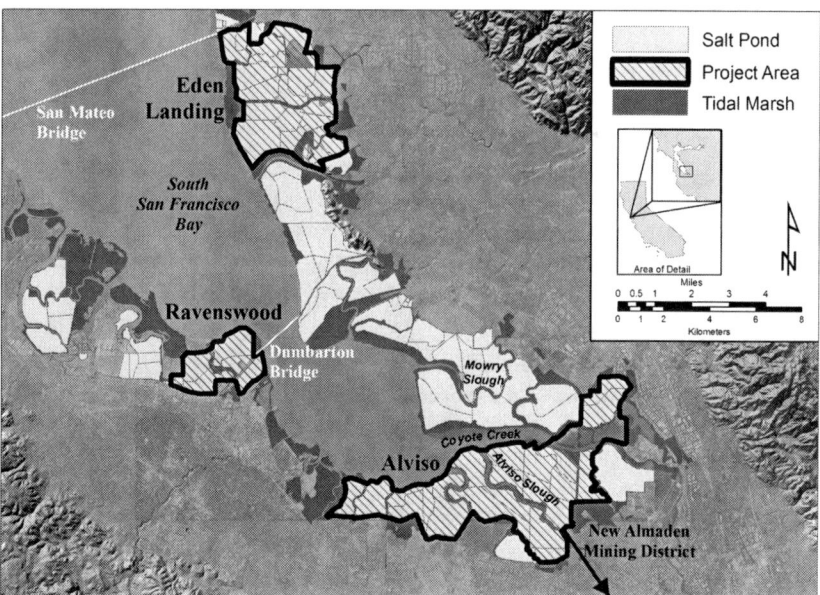

Fig. 3 The South Bay Salt Pond Restoration Project encompasses 15,100 A south of the San Mateo Bridge in San Francisco Bay. In this paper, South Bay is defined as the Bay south of the San Mateo Bridge, a subset of which is lower South Bay, defined here as the Bay south of Dumbarton Bridge

then. The first round of marsh restoration actions began with the breaching of a few ponds in 2006 and will continue with upcoming breaches planned in the next 2 years. Various alternatives for the final, restored South Bay landscape are under consideration, which differ mainly in location and in the ratio of restored marshes and managed ponds. Both habitat types are necessary to support endemic marsh wildlife and migratory water birds.

Both the importance of restoring the South Bay Salt Ponds and some of the associated challenges arise from the fact that the ponds are located in a highly urbanized Estuary. The Project is important for biological conservation, because so many tidal wetlands have been lost with the result that endemic wildlife are now endangered. One of the challenges for the Project is that, despite vast improvements in sewage treatment and water quality over the past several decades, the South Bay nevertheless faces substantial current and future pollution threats. Therefore, SBSPRP managers must consider water quality as a prime factor as they proceed to restore wetland habitats and manage the remaining ponds for wildlife. The location of the Project at the interface between land and water means that water quality in the Project will be tied to water quality management both in the South Bay watershed and in South San Francisco Bay itself. Furthermore, Project managers must be careful that restoration and associated actions do not exacerbate pollution problems in the region. In this paper, we explore the relationship between chemical contamination of the South Bay and the SBSPRP, including current conditions and how the Project and South Bay water quality may affect each other in the future.

2 Current Water Quality

Water quality in South San Francisco Bay is compromised by a variety of chemical contaminants and other types of pollutants that mainly originate in the watershed. In this paper, we focus on water quality in the Estuary, wherein most of the water quality research and monitoring has occurred, rather than on tributary water quality. Pollution in the Bay was present prior to the nascence of the SBSPRP and provides the water quality context in which the wetland restoration will proceed. Thus, the "before" restoration condition of the South Bay includes a history of legacy pollutants in the sediment and water that will affect the "after" restoration condition. In this pollution-impacted Estuary, the goal of the SBSPRP may be to maintain contaminants at or below current concentrations, rather than to restore habitats to pristine water quality.

2.1 Chemical Contaminants

Mercury, PCBs, and polybrominated diphenyl ethers (PBDEs) are the persistent contaminants of greatest concern in the South Bay. All three are present at elevated concentrations in both the abiotic environment and wildlife. Selenium, pyrethroid insecticides, polycyclic aromatic hydrocarbons (PAHs), and dioxins are also problematic. Legacy insecticides (DDTs, dieldrin, and chlordanes) have historically affected the Bay, and these remain above accepted thresholds of concern in a small proportion of samples. Brief summaries of the current state of knowledge for these contaminants in the San Francisco Estuary are given below, with an emphasis on recent data from the South Bay.

2.1.1 Mercury

Mercury is one of the primary current threats to water quality in the South Bay and tops the list of contaminant issues for the SBSPRP. In addition to the concern for present mercury concentrations, there is concern that restored wetlands could result in increased methylmercury production and bioaccumulation. This latter concern is addressed in Section 3.2.

There are several sources of total mercury contamination to the South Bay. These include legacy mercury from mercury- or gold-mining operations, respectively, in the Coast and Sierra Nevada ranges, and ongoing sources such as atmospheric deposition, runoff from urban and industrial areas, and outflow from wastewater treatment plants. The historic New Almaden mining district is situated in the hills above San Jose, in the watershed that drains through the Guadalupe River and Alviso Slough into lower South San Francisco Bay (Fig. 3). New Almaden was the largest mercury mine in North America. This mine ceased operations in 1976 when purchased by Santa Clara County. Legacy mercury from this mining district continues to enter the Bay today. Based on empirical measurements and a sediment-transport model, one load estimate for the Guadalupe River was 4–30 kg of mercury transported into lower South Bay per year (Thomas et al. 2002). More recent measurements indicate high inter-annual variation with much higher loads in peak years: 116, 15, and 8 kg in 2003, 2004, and 2005, respectively (McKee et al. 2006).

In a recent study, concentrations of total mercury and methylmercury were tracked in sediment from the SBSPRP ponds from 2003 to 2007 (Miles and Ricca 2010). Average total mercury in sediment in the Eden Landing ponds (0.11 $\mu g/g$) was below the South Bay ambient concentration (0.23–0.27 ppm) (SFEI 2009), whereas average concentrations exceeded the ambient level in the Alviso ponds (0.75–1.03 $\mu g/g$). Even within the Alviso pond complex, total mercury concentrations in sediment varied greatly among ponds. Sediment total mercury tended to be stable over time and was not correlated with methylmercury, which was more temporally variable (Miles and Ricca 2010). Understanding the variability, in space and time, of methylmercury concentrations in pond sediment was difficult, because seemingly similar ponds had different chemical and physical conditions that potentially affected mercury cycling. One notable change in methylmercury concentration was a greater than fivefold increase that occurred in two ponds after a levee breach returned them to tidal action. A third pond that was part of the same restoration action did not exhibit such a large increase in sediment methylmercury concentration (Miles and Ricca 2010).

Bioaccumulation of methylmercury also varies spatially across the SBSPRP area, but no long-term temporal trends are indicated by available data. In water birds, the highest exposure (as indicated by blood total mercury concentrations) was consistently observed in the western Alviso ponds, in both terns (Ackerman et al. 2008a) and recurvirostrids (Ackerman et al. 2007). The spatial pattern in marsh birds was completely different, however. Concentrations of total mercury in the blood of tidal marsh Song Sparrow (*Melospiza melodia pusillula*) were related to distance from the sampling site to the Bay, rather than to region within South Bay (Grenier unpublished data). Regarding time trends in methylmercury bioaccumulation, the best dataset available for the Estuary is for total mercury concentrations in striped bass muscle (*Morone saxatilis*) from 1970 to 2000. These data indicate no change over three decades (Fig. 4) (Greenfield et al. 2005). This result probably reflects the ongoing inputs to the Bay from both legacy and contemporary sources and flux from the massive reservoir of mercury in Bay sediments (Conaway et al. 2008).

Methylmercury bioaccumulates in South Bay food webs to concentrations that are sufficiently high to cause concern for adverse effects in humans and sensitive wildlife. Humans are at risk from exposure to methylmercury from eating sport fish from San Francisco Bay. The concentrations of mercury in sport fish taken from the Bay, and particularly from the South Bay, are higher than mercury residues in fish from other parts of California (Davis et al. 2007a). A consumption advisory for sport fish in the Bay was issued in 1994 and was driven by the concentrations of

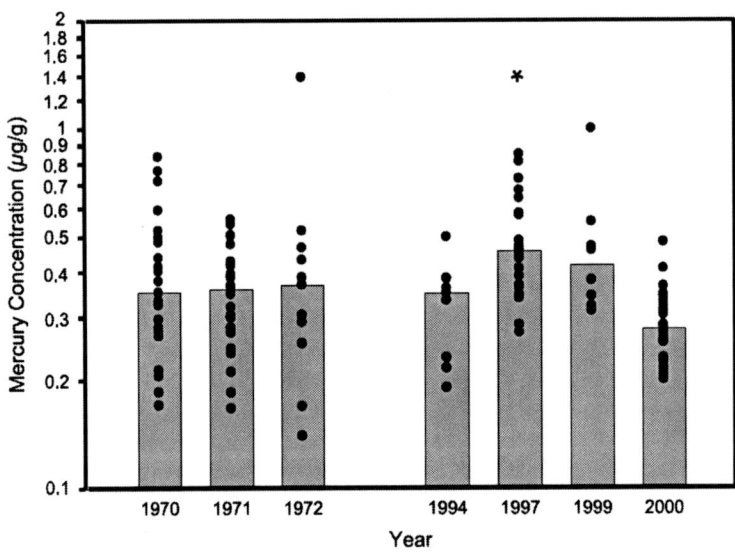

Fig. 4 Total mercury concentrations in striped bass muscle from San Francisco Bay showed no trend between 1970 and 2000. Gray bars indicate annual median concentrations. To correct for variation in fish length, all plotted data were calculated for a 55-cm fish using the residuals of a length:log (Hg) relationship. The asterisk above 1997 indicates significant difference from overall length:mercury regression (see original paper). Note log scale on the y-axis. From: Greenfield et al. (2005)

mercury, PCBs, organochlorine insecticides, and dioxins (OEHHA 1994). The public is advised not to eat Bay fish too frequently, and large predatory fish should not be consumed at all by certain demographic groups. Leopard shark (*Triakis semifasciata*), striped bass, and white sturgeon (*Acipenser transmontanus*) are examples of species that tend to have the highest mercury concentrations (Greenfield et al. 2005; Hunt et al. 2008).

Evidence of the exposure to and effects of methylmercury on wildlife in the Bay and its wetlands is increasing. Wildlife exposure to methylmercury is becoming better understood as the body of research has grown over the last 20 years; the effects of methylmercury on wildlife are less well studied. The effect thresholds of specific species constitute an important data gap, and linking elevated exposure to population-level effects is difficult. For example, a large proportion of Black-necked Stilt (*Himantopus mexicanus*) and Forster's Tern (*Sterna forsteri*) breeding adults had total blood mercury concentrations that placed them at moderate to high risk of reproductive impairment (Eagles-Smith et al. 2009). However, a related study of the effect of egg mercury concentrations on chick survival, in these same populations, failed to show significant effects (Ackerman et al. 2008b, c).

The overall change in habitat acreage from salt pond to tidal marsh and tidal slough that will occur as the SBSPRP progresses, and the particular locations of the ponds that are converted to marsh, will likely affect methylmercury concentrations in South Bay water birds. Where these birds forage, both in terms of region within the Bay and habitat type, is closely tied to the variability of mercury tissue concentrations of recurvirostrids, terns, and scaup (Ackerman et al. 2007, 2008a; Eagles-Smith et al. 2009). Water birds from the lower South Bay had the highest total mercury concentrations in their tissues (Eagles-Smith et al. 2009). The habitats that were associated with higher mercury concentrations in water birds were at the margin of the Bay, and these habitats (i.e., salt ponds and managed marsh) had altered hydrology (Ackerman et al. 2007, 2008a).

In contrast to the water birds discussed above, many tidal marsh bird species are endemic to San Francisco Bay. These tidal marsh birds also have elevated exposure to methylmercury and consequent health risks. In particular, the federally endangered California Clapper Rail has poor reproductive success that may be related to methylmercury. An estimated 15–30% of the observed reduction below normal hatchability in this subspecies has been attributed to contaminants, with methylmercury principal among them (Schwarzbach et al. 2006). Effects on other marsh birds in San Francisco Bay have not been studied, but information on the exposure of the Black Rail (*Laterallus jamaicensis*) (Tsao et al. 2009) and tidal marsh Song Sparrow (Grenier unpubl data) indicate that many breeding adults are above a 25% effect concentration (0.81 μg/g wet weight total mercury in blood) that is based on lab and field studies of other songbirds and above which 25% of eggs are predicted not to hatch due to methylmercury effects.

A recent egg-injection study provided information on the relative sensitivity of 26 bird species to methylmercury (Heinz et al. 2009) including representative species from the taxonomic families that have been studied in San Francisco Bay. Ducks (e.g., scaup and Mallard) had low sensitivity; terns, rails, and songbirds had medium sensitivity to mercury; and two ardeid (herons and egrets) species had high sensitivity, whereas another ardeid had medium sensitivity. This new information calls into question how the avian embryotoxic threshold – which is based on studies with captive Mallards (Heinz 1979) – should be interpreted for species that are more sensitive to mercury than Mallard. For example, wading-bird mercury exposure has not been studied for some time in South San Francisco Bay, but Black-crowned Night Heron (*Nycticorax nycticorax*) and Snowy Egret (*Egretta thula*) eggs collected between 1982 and 1990 (Ohlendorf et al. 1988; Hothem et al. 1995) had mean mercury concentrations that were less than half the Mallard threshold value. Given the difference in sensitivity of the ardeids and the Mallard, it is difficult to interpret whether the observed concentrations in South Bay wading birds are problematic or not.

Mammalian exposure and effects from methylmercury are less well characterized than they are for birds. Many mammalian wildlife populations that may have been sensitive to mercury are now extirpated from the Bay. The taxa of concern that remain in South Bay are harbor seals (*Phoca vitulina*) and small mammals in tidal marshes. Harbor seals in San Francisco Bay have elevated mercury concentrations in blood and hair (Kopec and Harvey 1995; Brookens et al. 2007), and Mowry Slough in the lower South Bay (Fig. 3) is an important breeding area for the seals. However, effect studies in harbor seals have indicated greater risk from organic contaminants than from mercury. Mercury exposure and effects in small mammals resident in tidal marshes are virtually unstudied. In the one study performed in San Francisco Bay, endangered salt marsh harvest mice were absent from marshes that had higher concentrations of mercury in other rodents, yet were present in marshes with lower rodent mercury residues (Clark et al. 1992). This broad correlation may indicate that high methylmercury bioaccumulation may be a stressor that contributes to extirpation of these mice in some marshes. A prime candidate for evaluation of mercury exposure in the tidal marsh ecosystem is the shrew. These tiny insectivores are completely carnivorous, have extremely high metabolic rates, and endemic marsh subspecies have been described for both ornate (*Sorex ornatus*) and wandering (*Sorex vagrans*) shrews in San Francisco Bay. Shrews are good bioindicators of metal pollution (Sanchez-Chardi et al. 2007), and in contaminated areas elsewhere shrews have accumulated liver mercury concentrations up to and in excess of 30 μg/g dry wt (Cocking et al. 1991; Talmage and Walton 1993).

2.1.2 PCBs

Despite the 1979 federal ban on PCB production and sale and subsequent gradual decline of PCBs in the environment, this suite of chemicals remains one of the main contaminants that affects water quality in San Francisco Bay, and the South Bay in particular. PCB residues are widely spread in Bay sediments, and they continue to bioaccumulate in the food web to a degree that poses health risks to humans and wildlife (Davis et al. 2007b). The most recent water quality monitoring data (SFEI 2009) as well as reviews of sediment PCB data for the Bay (Davis et al. 2007b; SFBRWQCB 2007) show that South Bay has relatively high concentrations, and PCB hotspots are present in the wetlands and other Bay margin habitats. The most important pathways by which PCBs enter South Bay waters are urban runoff and erosion of buried sediment. Riverine inputs from the Sacramento-San Joaquin Delta are important in other parts of San Francisco Bay but have less influence in South Bay, which is geographically and hydrologically more removed from the Delta (Conomos et al. 1979). Urban runoff is a source of PCBs that can be controlled and, as such, is a focus of the PCB total maximum daily load (TMDL) regulation. The effectiveness of different management options for reducing PCBs in stormwater is an important knowledge gap (Davis et al. 2007b). Release of PCBs through erosion of buried sediment is discussed later in this paper.

Along with mercury, PCBs are a primary driver of the advisory to limit consumption of fish taken from the Bay (OEHHA 1994). Median concentrations of PCBs in sport fish from South Bay have consistently exceeded the PCB TMDL cleanup target of 10 ng/g (Hunt et al. 2008). White croaker (*Genyonemus lineatus*) and shiner surfperch (*Cymatogaster aggregata*) are species with relatively high lipid content; these species had median PCB concentrations that exceeded the target by more than one order of magnitude. White sturgeon, anchovy (*Engraulis mordax*), and black surfperch (*Embiotoca jacksoni*) also consistently exceeded this value.

PCB concentrations in the Bay may be high enough to adversely affect wildlife, with fish-eating species at the top of the food web generally facing the greatest risks (Davis et al. 2007b). Since the early 1980s, the available data have indicated that PCBs accumulate to high concentrations in South Bay piscivorous birds and may cause adverse effects on survival. PCB concentrations in eggs well above 1.0 µg/g wet wt have been measured in Black-crowned Night Herons, Snowy Egrets, Forster's Terns, Caspian Terns, and Least Terns (*Sterna antillarum*) (Hoffman et al. 1986; Ohlendorf et al. 1988; She et al. 2008). Effects on growth and induction of cytochrome P450 were documented to have occurred in Black-crowned Night Herons and Double-crested Cormorants (Hoffman et al. 1986; Davis et al. 1997). The highest geometric mean concentrations in tern eggs were found in and near Eden Landing, one of the main restoration areas for the SBSPRP (Fig. 3; She et al. 2008). Even California Clapper Rails, a relatively low-trophic-level marsh species that consumes mainly invertebrates, were found to have some eggs with PCB concentrations sufficient to potentially cause deleterious effects (Schwarzbach et al. 2001). Authors of a recent study with a small number ($n = 4$) of fail-to-hatch Clapper Rail eggs reported a median PCB concentration in the eggs of 4,640 ng/g lipid wt (She et al. 2008). An obstacle to determining whether these concentrations may produce adverse affects is that the relative sensitivity of rails to PCBs is not known.

Harbor seals from San Francisco Bay, sampled in various studies over the past 30 years, had PCB concentrations in blubber and whole blood that exceeded those associated with impaired reproduction in a controlled feeding study (Davis et al. 2007b; Thompson et al. 2007). Total PCB concentrations in liver from five adult seals stranded in the Bay between 1989 and 1998 had a median concentration of 35.6 µg/g lipid wt (Park et al. 2009). Decreases in seal blood PCB concentrations occurred between the early 1990s and 2001–2002. Despite this reduction, some seals in the most recent time period may have experienced health effects related to organic chemical contaminants; this conclusion was based on a correlation between leukocyte counts and concentrations of PCBs, PBDEs, and DDE (Neale et al. 2005). Thus, PCBs are a significant concern in harbor seals.

Although the available data from top predators in the Bay were not collected with the intention of measuring long-term trends, there is a general pattern of slow PCB residue decline in wildlife over the past two to three decades (Davis et al. 2007b). The longer the time span of the dataset, the more apparent the decline, which indicates that the net dissipation of PCBs in the Bay food web is quite slow. The difference in analytical methods used to measure for PCBs over the years and variation in study designs and tissues sampled make this a tentative conclusion that will be clarified only with future monitoring.

2.1.3 PBDEs

Compared to mercury and PCBs, PBDEs have surfaced much more recently as contaminants of concern in San Francisco Bay. PBDEs are an example of emerging contaminants that are persistent and biomagnify, and that could affect higher-trophic-level species in restored habitats. PBDEs were virtually undetected in samples during the 1980s. However, over the course of the 1990s, PBDE residues became common in the water, sediments, and food web of the Bay. Such brominated flame retardants are relatively new contaminants of environmental concern, and the scientific understanding of their effects is limited (Birnbaum and Staskal 2004). Thresholds of concern for PBDEs have not yet been established. The two lower-brominated congener mixtures (penta-BDEs and octa-BDEs) were banned in California in 2006, but the deca-BDE mixture is still in commercial production.

Sources and pathways by which PBDEs enter the Bay are under investigation. Possible pathways include municipal and industrial discharges, stormwater, atmospheric deposition, small and large tributaries, and landfill leaching (Oram et al. 2008). Based on studies from tributaries and wastewater treatment plants in South Bay, PBDE loads are 3–10 times greater than those of PCBs (McKee et al. 2006; SFEI 2007). Monitoring data from 2002 to 2008 indicate that the lower South Bay is a hotspot for PBDEs in sediment and water, especially in certain years (Fig. 5; SFEI 2009). PBDEs are a high priority for ongoing monitoring and research to understand trends in food-web exposure and effects on sensitive species.

PBDE concentrations measured between 1989 and 2003 in harbor seal blubber, tern eggs, and human breast tissue from the San Francisco Bay Area were among the highest ever recorded (She et al. 2002, 2004). Some individual Forster's Tern eggs from the Eden Landing area of the SBSPRP (Fig. 3) had extremely high total PBDE concentrations of 62,400 and 63,300 ng/g lipid wt in 2001 and 2002, respectively (She et al. 2008). The geometric means for total PBDE concentrations in tern eggs sampled from 2000 to 2003 were lower, but still quite elevated – in the range of 3,700–4,800 ng/g lipid wt for Caspian, Forster's, and Least Terns (She et al. 2008). PBDEs have also been documented to exist in the sediment, water, bivalves, and fish of the Estuary (Holden et al. 2003; Oros et al. 2005). Mean concentrations (wet wt) in fish collected in 2006 were 56 ng/g for white croaker, 20 ng/g for white sturgeon, 13 ng/g for shiner surfperch, and 12 ng/g for northern anchovy (Hunt et al. 2008).

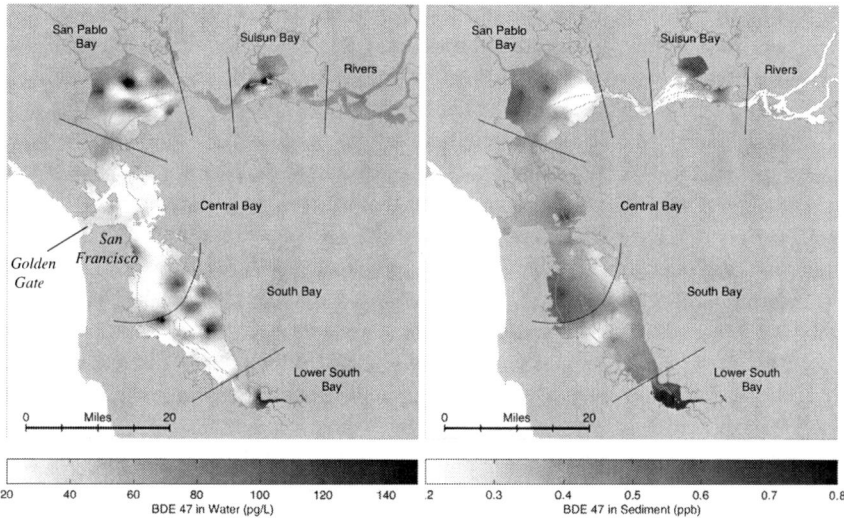

Fig. 5 Spatial trends of Brominated Diphenyl Ether (BDE) 47 in water from 2002 to 2008 (left) and in sediment from 2004 to 2008 (right) for the San Francisco Estuary. The two most southern segments of the Bay, as delineated between the black lines, correspond to the location of the South Bay Salt Pond Restoration Project (SBSPRP). These segments contain several hotspots for BDE 47, which is one of the most abundant Polybrominated Diphenyl Ethers (PBDEs) and is an index of PBDEs as a group. The highest average concentration from 2004 to 2008 of BDE 47 in sediment was 0.75 ppb in lower South Bay. From: SFEI (2009)

2.1.4 Other Chemical Contaminants

Polychlorinated dibenzodioxins and dibenzofurans are other persistent contaminants of concern in South Bay, and they are also among the pollutants in fish that prompted the fish consumption advisory for San Francisco Bay. The Regional Monitoring Program for Water Quality in the San Francisco Estuary has been analyzing residues of dioxins and dioxin-like compounds in fish (furans and coplanar PCBs) every 3 years, since 1994. The results have been relatively consistent throughout that time period, suggesting that dioxins are maintaining relatively constant concentrations in fish (Fairey et al. 1997; Davis et al. 2002; Greenfield et al. 2005; Hunt et al. 2008). Dibenzodioxins and dibenzofurans in all the fish (white croaker) sampled in 2000, 2003, and 2006, exceeded the screening value of 0.3 pg/g wet wt dioxin toxic equivalents (TEQs) established by the California Office of Environmental Health Hazard Assessment. Dioxin TEQ concentrations in 2006 were highest in the South Bay, with a maximum level of 16.3 pg/g wet wt (Hunt et al. 2008).

There is evidence that dioxins and dioxin-like compounds may be causing effects in two federally endangered bird species: the Least Tern and California Clapper Rail. Ten Least Tern eggs from the 2001 and 2002 breeding seasons had a geometric mean of TEQs that was within the effects threshold range calculated by the study authors (Adelsbach and Maurer 2007) using avian-specific toxic equivalency factors. These eggs had failed to hatch, so they represented a biased sample of the population. However, their failure to hatch could have been related to high TEQs. Clapper Rail fail-to-hatch eggs ($n = 4$) exhibited concentrations of individual dioxins above a threshold for effects in chickens (Adelsbach and Maurer 2007).

Once at the top of the list of contaminants of concern for San Francisco Bay, legacy insecticides are now declining. Historical use of dieldrin, chlordanes, and DDTs resulted in widespread contamination of the Bay and resident wildlife, and they were each banned 20 or more years ago. Legacy pesticides persist in Bay sediment and continue to enter the Bay from its tributaries (Connor et al. 2004), but the current situation is much improved. Reviews are available on the bioaccumulation behavior of these contaminants from statewide monitoring programs (Davis et al. 2007a) and San Francisco Estuary monitoring (Gunther et al. 1999; Greenfield et al. 2005). These reviews document significant residue declines of these contaminants in fish and bivalves over the past two to three decades. However, the rate of the decline may have slowed since the early 1990s, and the Bay may not be adequately cleansed of these chemicals until ongoing inputs are eliminated (Connor et al. 2004).

A small proportion of sport fish sampled from the Estuary exceed human health thresholds of concern for legacy insecticides. The most recent published data from 2006, when three fish species were sampled, show that most of the exceedances were for dieldrin (32%, 9 of 28 composites), and there were fewer for DDTs (11%, 3 of 28) and chlordanes (4%, 1 of 28) (Hunt et al. 2008). For each contaminant, some or all of the composites that exceeded the threshold of concern were from South Bay. The evidence suggests that current effects of legacy insecticides on wildlife are minimal. In some places, chlordanes may impact benthic invertebrates, but the impacts of DDTs on bird reproduction that were documented in the 1980s are no longer in evidence (Thompson et al. 2007).

A variety of other organic chemicals are cause for concern in the Bay, including pyrethroid insecticides and PAHs, but fewer data are available for these compounds than for the chemical groups discussed above. Pyrethroids are relatively new insecticides and are used as alternatives to the legacy organochlorine and organophosphate insecticides. As the use of pyrethroids as agricultural, commercial, and household insecticides increases, the potential for effects on invertebrates and fish, to which pyrethroids are highly toxic, also increases. Lowe et al. (2007) examined the toxicity of sediments in six tributary creeks around the Bay and found that only the South Bay tributaries were toxic to amphipods. The cause of this toxicity was identified as likely originating from exposure to either pyrethroids or DDT metabolites. Amweg et al. (2006) found that eight creeks along the eastern side of the Bay displayed toxicity from pyrethroids, mainly bifenthrin, on at least one of four sampling occasions. The effect of the pyrethroids on biota upon entering the Bay at the mouth of the creeks is not known.

PAHs are organic contaminants derived from carbon-based fuels, such as petroleum products. These chemicals are persistent in sediment and water, but they are metabolized by vertebrates, rather than accumulating in vertebrate tissue.

Monitoring studies suggest that PAH sediment concentrations may be above reproductive effects thresholds for fish (SFEI 2007). A new study has been funded by the Regional Monitoring Program to examine the effects of PAHs on larval fish in the Bay. PAH concentrations have been relatively constant over the past two decades (SFEI 2007). Events such as the Cosco Busan Oil Spill in November 2007 in central San Francisco Bay are a reminder that the high density of shipping traffic renders the Estuary vulnerable to PAH contamination from oil spills.

Besides mercury, other elemental contaminants that have been studied in the Bay include selenium, nickel, and copper. Selenium bioaccumulates in the San Francisco Bay food web to an extent that could cause harm to fish and to humans that consume diving ducks, but the focus of that concern is in the northern reaches of the Estuary. A consumption advisory for ducks is in place and a selenium TMDL is being developed. Nickel and copper were considered historical problems in San Francisco Bay, but samples obtained over the past decade indicate that concentrations are consistently lower than the water quality objective considered to be completely protective of aquatic life (SFEI 2007). Concentrations of arsenic, cadmium, chromium, copper, lead, nickel, selenium, and zinc, measured in the Project pond sediments close to the time of public acquisition, were generally at or below ambient concentrations in the surrounding area, although some ponds had above-ambient selenium levels (Brown and Caldwell 2005).

Despite the contaminant monitoring and research summarized above, many chemicals in the Bay may be causing unrecognized effects, and, conversely, some documented effects cannot be tied to specific contaminants. Evidence from biomarkers in fish and benthos, such as breaks in DNA and cellular abnormalities, shows that these taxa have been exposed to contaminants, yet pinning down which chemical has induced the effect is difficult in an Estuary with a complex cocktail of environmental pollutants. Toxicity of sediment to benthic invertebrates is prevalent around the margin of the Bay, yet the particular cause or causes have not been easy to identify. Contaminant mixtures may be as important, or more important, than individual contaminants (Thompson et al. 2007). A recent study in San Francisco and Tomales Bay tidal marshes documented benthic marsh fish population effects that correlated with a variety of contaminants in sediment (McGourty et al. 2009).

Emerging contaminants are a diverse group of unregulated and relatively unmonitored chemicals that have been detected in the environment and whose effects are largely unknown. Some chemicals considered to be emerging contaminants for the Estuary include perfluorinated chemicals (PFCs), pharmaceuticals, and non-PBDE flame retardants. Harbor seals recently analyzed for PFCs had concentrations in blood that were several times greater than those found in seals from other parts of the world and, seals from the South Bay had particularly high concentrations (SFEI unpubl data). Recent data from lower South Bay documented the presence of pharmaceuticals at concentrations well below available acute and chronic toxicity thresholds (SFEI unpubl data). Current use, non-PBDE flame retardant chemicals have also been detected in South Bay by the Regional Monitoring Program (unpubl data).

2.2 Other Aspects of Water Quality

2.2.1 Biological

Water quality in the Bay is also affected by biological constituents, particularly invasive species. San Francisco Bay is one of the most invaded estuaries on the planet, both in terms of the numbers of exotic organisms and their ecological dominance (Cohen and Carlton 1998). The arrival of exotic organisms in ballast water of ships and via other human activities has altered the ecology and food web of the Bay in ways so significant that water quality has also been profoundly affected. Hundreds of non-native taxa have been identified in Bay waters, and more than one hundred others exist that are of unknown origin (possibly non-native). These invaders dominate a large number of biological communities in the Bay, typically comprising 99% of the biomass of soft-bottom benthos, fouling communities, brackish-water zooplankton, and freshwater fish (Cohen and Carlton 1998). The Asian clam (*Corbula amurensis*) has radically altered the biological productivity of the northern subembayments by eliminating seasonal phytoplankton blooms that previously fueled flourishing populations of zooplankton, invertebrates, and fish in the pelagic food web (Thompson 2005). *Corbula* filtered nearly all of the phytoplankton from the water, thus suppressing the seasonal blooms. This effect of *Corbula* did not occur in South Bay, where differences in physical factors caused the phytoplankton bloom to occur earlier in the year, reducing the influence of this exotic clam (Thompson 2005). Nevertheless, the potential for exotic species to radically alter water quality clearly exists in South Bay.

Other changes in phytoplankton blooms have been occurring in San Francisco Bay and are thought to result from changing physical factors rather than invasive species. In 2004, the first dinoflagellate bloom, or red tide, observed in nearly 30 years of monitoring occurred in South San Francisco Bay (Cloern et al. 2005). These blooms can be toxic, causing die-offs of fish and other organisms, but in this case the bloom ended before any negative effects were observed. More recently, a cooling phase in the nearby Pacific Ocean was associated with various marine species moving into the Bay seeking warmer waters (Cloern et al. 2007). These new arrivals preyed on clams and reduced their numbers. This resulted in a release of phytoplankton populations that increased biomass year-round and allowed new autumn–winter blooms. These fluctuations at the base of the food web can have significant effects on fisheries and wildlife. The evidence suggests that the recent changes in phytoplankton were related to climatic events that extended well beyond the domain of the Estuary and its watershed (Cloern et al. 2007). Thus, future surprises of this kind can be expected to occur in association with climate cycles and global climate change.

2.2.2 Chemical

Chemical attributes of water quality, such as dissolved oxygen and salinity, are important to the success of the SBSPRP and the health of South Bay aquatic life.

High salinity and/or low oxygen levels can occur in the enclosed ponds in summer as water evaporates and algae decompose. Low dissolved oxygen then may cause fish and invertebrates to move upward into the oxygenated water layer near the surface, where they are more likely to be depredated by birds (Lonzarich and Smith 1997). If hypoxia becomes extreme, aquatic animals die. Shortly after the ponds were acquired for the purposes of restoration, there were some pond discharges into the Bay in 2004 that exceeded regulatory thresholds for dissolved oxygen. Those events provided important lessons for how to monitor the ponds. Monitoring practices have since changed, and the ponds are managed to maintain optimum water quality for wildlife, insofar as possible. The vast acreage of the Project makes this management and monitoring a large and costly task. The restoration of ponds to tidal marsh will eliminate both dissolved oxygen and salinity water quality concerns and the need to manage and monitor them. However, water quality will need to be managed and monitored in some ponds that will remain non-tidal and serve as foraging areas for water birds.

3 Potential Future Changes Related to Restoration

The spatial extent of the SBSPRP is so large that the Project could have regional effects on water quality of the South Bay that extend beyond the Project areas (Fig. 3). Particularly in the lower South Bay, south of the Dumbarton Bridge, the Project has the potential to strongly influence water quality at a regional scale, because water residence time is long (Conomos et al. 1979) and the acreage of the Project ponds is comparable to that of the open waters of the Bay. The effect of Project implementation could be to worsen, improve, or not affect the already impaired water quality in South Bay.

3.1 Erosion of Contaminants at Depth

Accelerated erosion of buried sediment is a potentially serious regional threat to South Bay water and sediment quality. Studies by the US Geological Survey (USGS) have shown that the South Bay (Foxgrover et al. 2004; Jaffe and Foxgrover 2006) and other parts of the Bay (Jaffe et al. 1998; Cappiella et al. 1999) experience fluctuating periods of erosion and of sedimentation, probably related to changes in sediment supply (McKee et al. 2002; Jaffe and Foxgrover 2006). Opening salt ponds to tidal action will create a new demand for sediment and will likely cause erosion of buried sediment in some areas. Such erosion could pose a risk with respect to recovery of the South Bay from legacy contamination, because the layers of sediment that would be unearthed are from earlier decades when the Bay was generally more contaminated (van Geen and Luoma 1999).

Bathymetric surveys conducted in 1931, 1956, 1983, and 2005 provided the basis for recent analyses of South Bay erosion and deposition (Foxgrover et al. 2004; Jaffe and Foxgrover 2006). From 1931 to 1956, a period with rapid urbanization, industrialization, and little wastewater treatment, the South Bay experienced widespread deposition of relatively contaminated sediment. From 1956 to 1983, a period including an era of peak contamination in the 1960s and marked improvements with the onset of wastewater treatment in the 1960s and 1970s, the South Bay experienced net erosion. In the most recent time period, net deposition has once again occurred. The erosion and deposition varies by locale, with more erosion in the northern part of South Bay and more deposition in lower South Bay. These long-term patterns are a critical piece of information needed to predict the rate of improvement of Bay water quality in future decades.

Sediment coring studies around the Bay consistently show greater concentrations of legacy contaminants at depth than at the surface. Cores taken from a South Bay tidal marsh along Coyote Creek (Fig. 3) exemplify the pattern, with mercury peaking at a depth that corresponds to deposition in the mid-twentieth century (Fig. 6) (Conaway et al. 2004). Time lines for the cores were established using radiocarbon and pollen of introduced plants. This coring study showed that pre-mining concentrations of mercury in South Bay sediment were similar to those in other parts of the Bay from the same time period. This similarity indicates that natural weathering

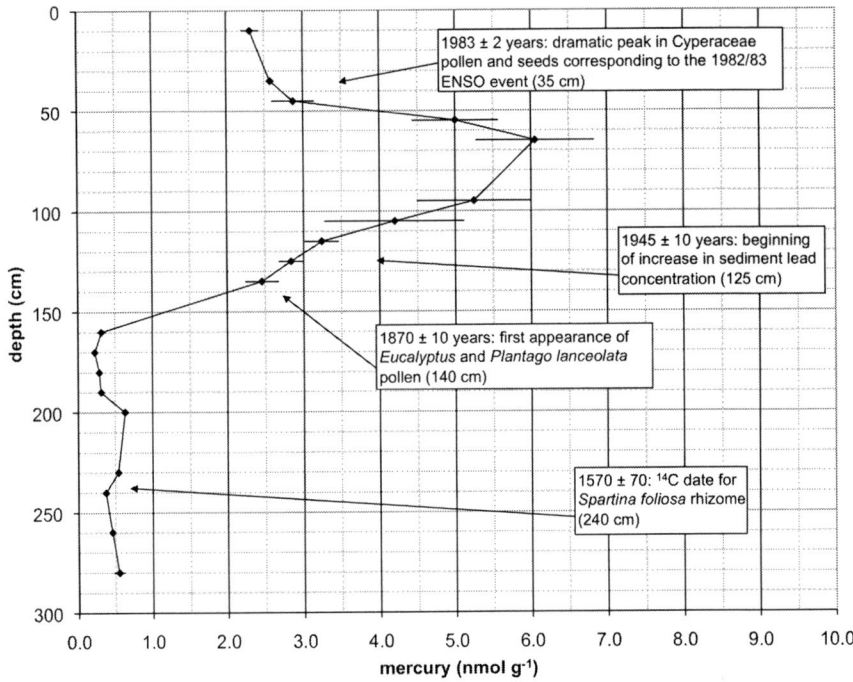

Fig. 6 In this sediment core from South Bay (Triangle Marsh), total mercury concentrations peaked at a depth corresponding to deposition in the mid-1900s. From: Conaway et al. (2004)

of mercury sources did not cause elevated total mercury in San Francisco Bay prior to mining; rather, the concentrations only increased after mining began. Pre-mining concentrations (0.08 µg/g) (Conaway et al. 2004) were approximately one-third of current concentrations, which average 0.24 µg/g (SFEI 2009).

A series of 15 sediment cores 2-m deep were taken from Alviso Slough, which currently drains the New Almaden district. These cores showed somewhat greater mercury concentrations at depth, compared to surface concentrations, and only a few cores with much higher mercury maxima below the surface (Marvin-DiPasquale and Cox 2007). Profiles of mercury by depth varied among the cores, and the subsurface concentration maxima were generally four to five times greater than the surface concentrations. This coring study included an experiment designed to assess whether mercury from deep sediment would become bioavailable if released into the water column by erosion. When buried sediment was mixed with oxygenated overlying water, reactive mercury ($Hg(II)_R$) concentrations increased 35- or 53-fold over 1 week, depending on salinity. Reactive mercury was operationally defined in this study as the fraction of mercury that was readily reduced to elemental Hg^0 by an excess of tin chloride ($SnCl_2$) over a short exposure time. Reactive mercury is thought to be available for conversion to methylmercury by sulfate-reducing bacteria. This result, and the increase in sediment MeHg (previously discussed) at two ponds after being returned to tidal action (Miles and Ricca 2010), suggests that a pulse of bioavailable mercury may be introduced to South Bay waters during erosion events associated with restoration actions. The duration of this pulse would probably be related to the duration of the erosion, which might last from months to years, depending on the restoration event. Modeling and monitoring of sediment erosion, transport, and fate processes, in conjunction with methylmercury monitoring, will provide important information for better understanding these relationships and the consequences of erosion in South Bay.

The data available for organic contaminants at depth in South Bay sediments are for PCBs and legacy pesticides, and these data also show higher concentrations below the surface. A subset of the cores collected from Alviso Slough (Marvin-DiPasquale and Cox 2007) was analyzed for PCBs and legacy pesticides (SCVWD 2008). PCB concentrations were consistently higher in the lower half, as compared to the upper half, in these 2-m cores. PCB concentrations in the upper half (52 ± 17 ng/g, mean ± st dev) exceeded a criterion (22 ng/g) set by the San Francisco Bay Regional Water Quality Control Board for the beneficial reuse of dredged material as wetland surface fill. PCB concentrations in sediment at depth (173 ± 17 ng/g) were similar to the criterion value for wetland foundation fill (180 ng/g). These criteria were used for comparison, since much of the scoured sediment will ultimately probably settle on extant tidal marsh and sloughs or in the breached ponds, which are the future sites of restored wetlands. PCB concentrations at the surface were similar to PCB values from randomly sampled nearshore surface sediments in San Francisco Bay (areas south of San Pablo Bay) (AMS 2005). PCB concentrations at depth were similar to those from sediment cores collected downstream of storm drains known to be contaminated by PCBs elsewhere in the Bay (AMS 2004).

Patterns of DDT concentrations were similar to those described for PCBs above, with one exception: DDT concentrations from the Alviso cores near the surface (27 ± 5 ng/g) and at depth (48 ± 16 ng/g) were higher than typical Bay concentrations (SCVWD 2008). Chlordane concentrations were 3.7 ± 2.3 ng/g at the surface and 16 ± 9 ng/g at depth. Concentrations in samples from the lower half of the cores exceeded the wetland foundation criterion (5 ng/g) and were above those found under ambient Bay conditions. Dieldrin was more concentrated in the lower half of the cores (2.0 ± 0.2 ng/g), and only the deeper samples exceeded the wetland surface screening guideline (0.7 ng/g) and were higher than typical Bay concentrations.

The contaminant depth profiles in sediment cores from South Bay exhibited patterns similar to those in cores from the northern reach of San Francisco Bay that were extensively studied for many contaminants and dated by the USGS (van Geen and Luoma 1999). The authors found that metals (Hg, Pb, Cu, Zn, Ag), DDTs, PCBs, and PAHs all had low baseline concentrations prior to industrialization, from which concentrations increased during the early to mid-twentieth century (Hornberger et al. 1999; Pereira et al. 1999; Venkatesan et al. 1999). Most of the contaminants peaked at depth and then declined toward the surface, but PAHs, Cu, and Zn did not decline in more recent sediments. Similarly, concentrations of trace metals from anthropogenic sources peaked at depth in a sediment core from a North Bay tidal marsh (Hwang et al. 2009).

Remobilization of buried sediments that are more contaminated than surface sediments poses a significant issue for restoration activities. Prior research has suggested that legacy contaminants persist in the upper layers of Bay sediment for decades, because the top 30 cm stay in the active layer due to mixing (Fuller et al. 1999). Both organic chemical contaminants and mercury from the Alviso cores showed a pattern of increasing concentrations at depth, necessitating consideration of how erosion could result in increased concentrations of these contaminants in the food web.

3.2 Change of Habitat Types

The planned restoration of salt ponds to tidal marsh has raised concerns about the possibility of increased net methylmercury production and subsequent accumulation in the food web. This concern applies not only to the restored marshes, but also to the South Bay as a whole, which could be affected at a regional scale. These concerns are based on studies in freshwater wetlands conducted at different locations in the USA. Local studies are underway to assess mercury risks associated with restoring particular ponds in South San Francisco Bay. An additional contaminant issue regarding the transformation of ponds to tidal marsh is that the conversion will involve sequestration of millions of cubic meters of sediment. The key question about sediment sequestration in marshes relates to whether it will remove contaminated sediment from the active sediment layer of the Bay and whether it will create marshes that have contaminated food webs.

3.2.1 Methylmercury Production in Wetlands

Wetland biogeochemical conditions may be conducive to the production of methylmercury. Sulfate-reducing bacteria are abundant in wetlands as a result of the anaerobic conditions that prevail in these organic-rich environments, and these bacteria are the main agents of mercury methylation. Many wetlands also have peat soils, and sediment that retains a high percentage of organic matter has been correlated with high concentrations of methylmercury (Krabbenhoft et al. 1999).

The central concern is that the restored marshes may cause greater accumulation of methylmercury in the food web than is already present. This concern has arisen from several studies performed in various parts of the USA. Results from these studies have linked wetlands to relatively high net methylmercury production and export into waterways (Selvendiran et al. 2008). Local- and regional-scale studies have correlated wetland acreage in upstream watersheds with high rates of methylmercury production and export in Wisconsin rivers (Hurley et al. 1995) and have identified wetlands as methylmercury sources in boreal forest (St. Louis et al. 1996) and in riparian wetlands of Massachusetts (Waldron et al. 2000). Additionally, a national-scale study of mercury contamination along multiple gradients found that sediment methylmercury contamination was most strongly correlated with proportion of wetland in the sub-basin (Fig. 7) (Krabbenhoft et al. 1999).

More limited research on this topic has been completed in San Francisco Bay, although some studies are ongoing. The results published to date suggest that wetlands are sites of variable and sometimes high rates of net methylation. Sediment methylmercury concentrations have been observed to be higher in tidal marsh (3–5 ng/g dry wt) than in the open waters of North San Francisco Bay (0.7 ng/g), despite having similar total mercury concentrations (Marvin DiPasquale et al. 2003). Krabbenhoft et al. (1999) found that methylmercury in sediments of five

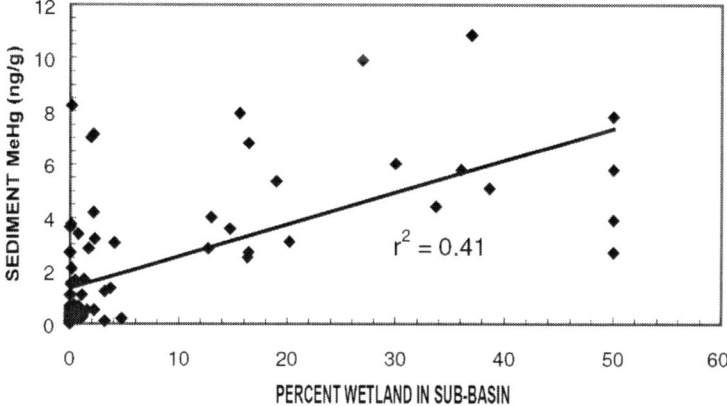

Fig. 7 Methylmercury concentrations in sediment were positively correlated to the percent of wetlands in the sub-basin in a national-scale study of 106 sites from 21 basins. From: Krabbenhoft et al. (1999)

tributaries of the Sacramento-San Joaquin Delta ranged from 0.55 to 2.84 ng/g, with methylmercury accounting for 0.1–2.2% of total mercury. The average percent of methylmercury present for these five sites was the lowest observed in 21 study basins across the USA. In contrast, in a 1994 investigation of tidal marsh sediments in the Bay, Schwarzbach et al. (2000) found methylmercury concentrations between 0.41 and 25.2 ng/g, with percent methylmercury comprising 0.1–6.6%. The maximum value of percent methylmercury to total mercury falls above the 80th percentile of data from the 21 study basins mentioned above, but the minimum value is quite low. Because free methylmercury is short-lived, percent methylmercury is considered to be an indicator of net methylmercury production (Gilmour et al. 1998). The high maximum percent of methylmercury that has been measured in Bay tidal wetlands indicates the potential for high net methylation rates in these ecosystems, while the low minimum percent methylmercury shows that not all tidal marshes are problematic. Understanding the underlying wetland characteristics and processes that lead to this variation is a key consideration for the SBSPRP and other wetland projects that seek to minimize mercury risks related to restoration actions.

Hydrology and organic matter may be the key factors that govern net methylmercury production in Bay wetlands. Research from other parts of the country indicates that hydrology is important in determining which wetlands have higher methylmercury production. St. Louis et al. (1996) found that spatial heterogeneity among wetlands in export of methylmercury was related to differences in hydrology. Newly flooded areas have been linked to spikes of methylmercury concentrations in water and the food web that can continue at elevated levels for decades (Bodaly et al. 2007). Experimental flooding of a boreal forest wetland caused it to increase in methylmercury export by a factor of nearly 40 (Kelly et al. 1997). The applicability of this result to tidal wetlands is uncertain, as some of the change after flooding the forest wetland was related to the death and decomposition of stressed vegetation (Kelly et al. 1997). In contrast, tidal marsh vegetation is adapted to tidal flooding; therefore, a flooded marsh may not be an appropriate parallel to a wetland where plants die after flooding. However, tidal wetlands in Louisiana were found to be probable sites of high methylmercury production and likely sources of methylmercury contamination in local marine food webs (Hall et al. 2008).

Results from San Francisco Bay tidal wetlands do point toward hydrology and organic matter as important factors relating to methylmercury production. A study comparing sediment methylmercury and methylmercury:total-mercury ratios between marsh interior and marsh edge showed concentrations of both to be higher at interior sites (Heim et al. 2007). Physical processes of tide and sediment transport govern marsh geomorphology, generally resulting in oxygenated, inorganic marsh edge sediments near the banks of tidal creeks and more poorly drained, peaty sediments in the marsh interior (Collins et al. 1986; Collins and Grossinger 2004; Culberson et al. 2004). Also, experimental work (Windham-Myers et al. 2009) has shown that methylmercury production in sediment decreased in some tidal wetlands of San Francisco Bay when plants were removed. The authors hypothesized that the mechanism for this reduction was a reduction in labile carbon (acetate) exuded by the plant roots that was previously fueling microbial activity and, hence, mercury methylation.

3.2.2 Sequestration of Contaminants in Marshes

Wetlands sequester and break down some contaminants, largely through processes that take place in sediment and secondarily in plants (Gambrell 1994; Reddy et al. 1999). Wetlands are such effective tools for remediation of pollutants that they are often constructed with the purpose of improving water quality, particularly for wastewater (Hammer and Bastian 1989; Reddy and d'Angelo 1997; Sheoran and Sheoran 2006). The restored South Bay tidal marshes will be created by the deposition in salt ponds of vast amounts of sediment with contamination at levels either of the ambient Bay or of local tributaries that influence particular sites. Restored marshes will begin to vegetate with cordgrass at low-marsh elevations, and they will continue to accrete sediment and associated contaminants as they age, until reaching a maximum elevation of many centimeters above mean high water (Collins et al. 1986; Collins and Grossinger 2004). Thus, the South Bay restored marshes may be sinks for persistent, sediment-bound contaminants. Although the restored wetlands may act as sinks for contaminated sediments, whether the net effect on Bay water quality will be beneficial is difficult to predict, given the potential remobilization of more contaminated sediment layers through erosion caused by restoration.

Contaminants in marsh surface sediments would likely bioaccumulate in primary producers and biota higher in the food web. Creating habitat for wildlife is one of the main goals of the SBSBRP, including habitat for endangered species that are marsh obligates. Thus, it will be important for the restored marshes to provide relatively uncontaminated food resources for species such as the California Clapper Rail. The magnitude of any contaminant effects on wildlife will depend partly on the degree of contamination in the sediment the marshes accrete and in the tidal water. Thus, the health of marsh biota will depend in part on the state of water and sediment quality in the Bay at the time the marshes begin to vegetate to become highly productive habitats capable of attracting dense wildlife populations. The salt ponds are to be restored in phases that, at any point in time, will yield restored marshes at different stages of development. Some ponds may not reach water elevations that will allow vegetation to take hold for decades, either because of low initial elevation, insufficient sediment supply, or sea level rise. Thus, any reductions in contaminant sources to the South Bay over the coming years will benefit the SBSPRP, and any increases will be detrimental to the Project.

3.2.3 The Effect of Bird Populations on Water Quality in Managed Ponds

Some of the former salt ponds will not be restored to tidal action and, instead, will be managed to support migratory water birds in winter and resident breeding birds in summer. The Pacific Flyway populations of water birds that migrate through the South Bay will encounter less managed pond acreage and more marsh acreage over time as the SBSPRP changes the landscape. The restored marshes will provide habitat for some species in the form of pannes (unvegetated tidal ponds in the marsh) and sloughs (tidal channels), yet many birds will continue to rely on the resources of the remaining managed ponds to build up fat reserves during winter and to support reproduction during the breeding season. These individuals may be crowded by necessity into dense aggregations in the remaining managed ponds.

The effect of dense bird populations on water quality is variable, with the outcome depending on the details of the situation. Bird feces can degrade water quality through eutrophication or introduction of pathogens. Some studies of eutrophication of water bodies by avian populations indicate that water birds contributed a significant proportion of total nutrients and degraded water quality (Manny et al. 1994). In one investigation of a shallow urban lake, the bird feces comprised nearly all the phosphorus loading (Scherer et al. 1995). Large bird populations can also introduce substantial quantities of bacteria to water bodies from their droppings (Valiela et al. 1991; Levesque et al. 1993; Graczyk et al. 2008), although in some studies water quality was not affected by bacteria from bird feces (Levesque et al. 2000). The body size of the bird, the density of the population, and the ability of the water body to dilute avian feces are important considerations in determining whether water quality might be impacted.

In a recent study of two managed ponds in the SBSPRP, it was found that the relationship between bird use and water quality was seasonal and related to how pond hydrology was managed. Shellenbarger et al. (2008) found that fecal indicator bacteria concentrations in managed ponds were higher in summer than in winter, despite bird abundance on the ponds being 10 times greater in winter. The researchers concluded that, in the summer, water from an adjacent slough with poor microbial water quality entered the ponds and increased concentrations of these bacteria. In the winter, bird feces probably contributed large quantities of fecal indicator bacteria. Whether this pattern holds true for other ponds in the SBSPRP will depend on factors such as hydrology, intensity of bird use, and quality of the source water to the ponds.

4 Future Changes that May Affect Water Quality

Other future changes that could affect water quality will include some changes that are beyond the control of the SBSPRP managers. Located in the transition area between the upland watersheds and the Bay, the Project has an intimate connection to the water quality of both areas. Increasing urbanization and human population in the San Francisco Bay Area and the sea level rise that is accompanying global climate change are likely to significantly impact the waters of the Bay and of the Project. The human population of the Bay Area is expected to rise by 2 million over the next 30 years, which is an increase to 140% of the current population (ABAG 2007). Given that human activities have caused nearly all the water quality impairment that is now observed in San Francisco Bay, the advent of 40% more people in the local watersheds may adversely affect water quality, despite efforts to minimize their impact. New contaminants will continue to emerge, such as pesticides, pharmaceuticals, flame retardants, and nanotechnology waste. The effects of population growth may also include greater peak loads of storm water, greater absolute quantity of contaminants in storm water, greater volume of treated wastewater, more atmospheric loading of contaminants from vehicle exhaust, and more trash. Many strategies are being developed to counteract these effects of population growth, including better management and recycling of storm water, reduction of pollutant sources, reuse of wastewater, and better trash management.

Climate change is also likely to affect the SBSPRP, and changing precipitation patterns may be the aspect of climate change that most affects water quality in the Project. Pulses of water are predicted to come from the Sierra Nevada mountains earlier in the year, in winter/spring instead of spring/summer, due to earlier snowmelt and to precipitation falling as rain instead of snow (Miller et al 2003; Dettinger et al. 2004; Vanrheenen et al. 2004). This change would probably result in greater flooding in spring and larger loads of contaminated sediment being carried to the Bay during these events. Wetter springs are predicted to be followed by drier summers, which will reduce natural dry-season inflows to the Bay. Therefore, it is likely that dry-season flows will have a greater concentration of contaminants (less total water volume) and will be more dominated by wastewater than in the past.

5 Recommendations

A general recommendation for the SBSPRP managers, and for others managing wetland restoration at a regional scale, is to practice adaptive management and ongoing monitoring for water quality, particularly bioaccumulation of contaminants in the food web. The four main questions that need to be answered with adaptive management and ongoing monitoring are as follows:

- What are the present levels of contamination in locations to be restored and in adjacent habitats?
- What is the effect of different types of restoration on contaminant exposure of wildlife in restored areas and adjacent habitats?
- What restoration approaches in terms of habitat type, location, water management, etc., can minimize the bioaccumulation of contaminants in the food web locally and regionally?
- What is the effect of restoration on the South Bay sediment budget and long-term trends in South Bay regional contamination?

We recommend an approach to address these uncertainties that includes the following elements.

1. Most importantly, a long-term regional program of monitoring and research is needed that assesses contaminants prior to, during, and after each restoration action within the larger SBSPRP. Emphasis should be placed on exposure and effects in biota, along with ongoing synthesis of the information obtained into conceptual and numeric models that describe contaminant dynamics at local and regional scales.

Long-term monitoring should be performed to ascertain the impact of restoration actions on water quality relative to ambient condition on local and regional scales. This monitoring should include sampling of concentrations in sport fish as an index of human exposure and of marsh, pond, and Bay wildlife that are appropriate indices of exposure in the food webs of each habitat. Monitoring of other food-web or ecosystem components may also be useful in establishing long-term trends and spatial patterns, and biosentinel organisms with high-site fidelity will be useful in differentiating relative mercury bioavailability at fine spatial and temporal scales. Water quality monitoring should be conducted in all of the habitat types that are part of each restoration plan, and tidal marsh, managed pond, and intertidal mudflat habitats will be of particular interest to monitor with respect to methylmercury.

Detailed surveys should precede individual pond restoration projects to document existing concentrations of mercury and other contaminants and to evaluate the potential for increased food-web accumulation. Determining the impact that a restoration project has will depend on the availability of baseline information collected prior to the project start. For restoration projects likely to mobilize or erode large quantities of sediment, preliminary studies are needed to evaluate contaminant concentrations in surface and buried sediments and in the water and sediment supply. The effect of remobilized contaminated sediment accumulating in restored areas and adjacent habitats must be monitored.

Long-term monitoring of other water quality indicators will also be needed. To ensure that contaminants do not interfere with the health of wildlife in restored habitats, the Project will need information from toxicity testing to assess the effects of current-use pesticides and other non-persistent contaminants and to ascertain general trends in contaminant concentrations in South Bay.

Studies performed in association with restoration should contribute to development of a conceptual understanding of mercury cycling in Bay wetlands that allows prediction of bioaccumulation in restored habitats, including different sub-habitats within wetlands. Food-web monitoring should be coupled with strategic process studies crafted to disclose the mechanisms of variation in methylmercury bioaccumulation within and among tidal wetlands. This knowledge will provide the foundation for environmental managers and engineers to develop designs that minimize the impact of restoration activities. High priority should be given to examining the effects of restoration on bioavailability and net methylation rates, as these processes have the potential to increase methylmercury exposure in biota.

Conceptual and numeric models of contaminant fate are required on local and regional scales. These models provide a framework for organizing the current state of knowledge and for defining uncertainties, and they should continue to be updated with new information.

2. Studies are needed that provide better information on the sensitivity of species facing the greatest exposure to methylmercury and other contaminants.

More information on sensitivity to methylmercury of California Clapper Rails, terns, and harbor seals is a priority. Piscivorous species are also highly exposed to

PCBs, dioxins, PBDEs, and other persistent organic chemicals, yet the sensitivity of these species to these individual chemicals and combinations of them is not well known.

3. Development of a sediment-transport dynamics model and sediment budget is required that accurately describe sediment mixing, deposition, and erosion in South Bay.

Numerical models are needed to predict the sources and quality of sediment that are supplied to restored wetlands, and the impacts of the Project on sediment erosion and possible contaminant remobilization at a regional scale.

6 Summary

The SBSPRP is an extensive tidal wetland restoration project that is underway at the margin of South San Francisco Bay, California. The Project, which aims to restore former salt ponds to tidal marsh and manage other ponds for water bird support, is taking place in the context of a highly urbanized watershed and an Estuary already impacted by chemical contaminants. There is an intimate relationship between water quality in the watershed, the Bay, and the transitional wetland areas where the Project is located. The Project seeks to restore habitat for endangered and endemic species and to provide recreational opportunities for people. Therefore, water quality and bioaccumulation of contaminants in fish and wildlife is an important concern for the success of the Project.

Mercury, PCBs, and PBDEs are the persistent contaminants of greatest concern in the region. All of these contaminants are present at elevated concentrations both in the abiotic environment and in wildlife. Dioxins, pyrethroids, PAHs, and selenium are also problematic. Organochlorine insecticides have historically impacted the Bay, and they remain above thresholds for concern in a small proportion of samples. Emerging contaminants, such as PFCs and non-PBDE flame retardants, are also an important water quality issue. Beyond chemical pollutants, other concerns for water quality in South San Francisco Bay exist, and include biological constituents, especially invasive species, and chemical attributes, such as dissolved oxygen and salinity.

Future changes, both from within the Project and from the Bay and watershed, are likely to influence water quality in the region. Project actions to restore wetlands could worsen, improve, or not affect the already impaired water quality in South Bay. Accelerated erosion of buried sediment as a consequence of Project restoration actions is a potentially serious regional threat to South Bay water and sediment quality. Furthermore, the planned restoration of salt ponds to tidal marsh has raised concerns about possible increased net production of methylmercury and its subsequent accumulation in the food web. This concern applies not only to the restored marshes, but also to the South Bay as a whole, which could be affected on a regional scale. The ponds that are converted to tidal marsh will sequester millions of cubic meters of sediment. Sequestration of sediment in marshes could remove contaminated sediment from the active zone of the Bay but could also create marshes with contaminated food webs. Some of the ponds will not be restored to marsh but will be managed for use by water birds. Therefore, the effect of dense avian populations on eutrophication and the introduction of pathogens should be considered. Water quality in the Project also could be affected by external changes, such as human population growth and climate change.

To address these many concerns related to water quality, the SBSPRP managers, and others faced with management of wetland restoration at a regional scale, should practice adaptive management and ongoing monitoring for water quality, particularly monitoring bioaccumulation of contaminants in the food web.

References

ABAG (2007) Projections 2007: City, county and census tract forecasts 2000–2035. Association of Bay Area Governments, Oakland CA

Ackerman JT, Eagles-Smith CA, Takekawa JY, Demers SA, Adelsbach TL, Bluso JD, Miles AK, Warnock N, Suchanek TH, Schwarzbach SE (2007) Mercury concentrations and space use of pre-breeding American avocets and black-necked stilts in San Francisco Bay. Sci Tot Environ 384:452–466

Ackerman JT, Eagles-Smith CA, Takekawa JY, Bluso JD, Adelsbach TL (2008a) Mercury concentrations in blood and feathers of pre-breeding Forster's terns in relation to space use of San Francisco Bay habitats. Environ Toxicol Chem 27:897–908

Ackerman JT, Eagles-Smith CA, Takekawa JY, Iverson SA (2008b) Survival of postfledging Forster's terns in relation to mercury exposure in San Francisco Bay. Ecotoxicol 17: 789–801

Ackerman JT, Takekawa JY, Eagles-Smith CA, Iverson SA (2008c) Mercury contamination and effects on survival of American avocet and black-necked stilt chicks in San Francisco Bay. Ecotoxicol 17:103–116

Adelsbach TL, Maurer T (2007) Dioxin toxic equivalents, PCBs, and PBDEs in eggs of avian wildlife of San Francisco Bay. USFWS, California/Nevada Operations, Sacramento, CA

Amweg EL, Weston DP, You J, Lydy MJ (2006) Pyrethroid insecticides and sediment toxicity in urban creeks from California and Tennessee. Environ Sci Tech 40:1700–1706

AMS (2004) Analysis of pollutants in sediment cores near storm water inputs – final report. Prepared by Applied Marine Sciences for the San Francisco Bay Clean Estuary Partnership, Oakland, California, July, 2004. www.cleanestuary.com

AMS (2005) Existing data on PCB concentrations of nearshore sediments and assessment of data quality. Prepared by applied marine sciences for the clean estuary partnership, Oakland, CA. www.cleanestuary.com

Birnbaum LS, Staskal DF (2004) Brominated flame retardants: Cause for concern? Environ Health Perspect 112:9–17

Bodaly R, Jansen W, Majewski A, Fudge R, Strange N, Derksen A, Green D (2007) Postimpoundment time course of increased mercury concentrations in fish in hydroelectric reservoirs of northern Manitoba, Canada. Arch Environ Contam Toxicol 53:379–389

Brookens TJ, Harvey JT, O'Hara TM (2007) Trace element concentrations in the Pacific harbor seal (*Phoca vitulina richardii*) in central and northern California. Sci Total Environ 372:676–692

Brown and Caldwell (2005) South Bay Salt Pond Restoration Project. Water and Sediment Quality Existing Conditions Report, 59 pp. http://southbayrestoration.org/pdf_files/Water_and_Sed_Quality_Existing_Conditions.3.30.05.pdf. Accessed on 3 December 2009.

Cappiella K, Malzone C, Smith R, Jaffe B (1999) Sedimentation and bathymetry changes in Suisun Bay: 1867–1990. U.S. Geological Survey Open-File Report 99-563. U.S. Geological Survey, Menlo Park, CA

Clark DR Jr, Foerster KS, Marn CM, Hothem RL (1992) Uptake of environmental contaminants by small mammals in pickleweed habitats at San Francisco Bay, California. Arch Environ Contam Toxicol 22:389–396

Cloern JE, Schraga TS, Lopez CB, Knowles N, Grover Labiosa R, Dugdale R (2005) Climate anomalies generate an exceptional dinoflagellate bloom in San Francisco Bay. Geophys Res Lett 32, L14608, doi:10.1029/2005GL023321

Cloern JE, Jassby AD, Thompson JK, Hieb KA (2007) A cold phase of the East Pacific triggers new phytoplankton blooms in San Francisco Bay. Proc Natl Acad Sci 104:18561–18565

Cocking D, Hayes R, King ML, Rohrer MG, Thomas R, Ward D (1991) Compartmentalization of mercury in biotic components of terrestrial flood plain ecosystems adjacent to the South River at Waynesboro, VA. Water Air Soil Pollut 57:159–170

Cohen AN, Carlton JT (1998) Accelerating invasion rate in a highly invaded estuary. Science 279:555–557

Collins JN, Grossinger RM (2004) Synthesis of scientific knowledge concerning estuarine landscapes and related habitats of the South Bay Ecosystem. Technical report of the South Bay Salt Pond Restoration Project. San Francisco Estuary Institute, Oakland, CA, 92 pp

Collins LM, Collins JN, Leopold LB (1986) Geomorphic processes of an estuarine marsh: Preliminary results and hypotheses. Int Geomorph 1:1049–1072

Conaway CH, Watson EB, Flanders JR, Flegal AR (2004) Mercury deposition in a tidal marsh of south San Francisco Bay downstream of the historic New Almaden mining district, California. Mar Chem 90:175–184

Conaway CH, Black FJ, Grieb TM, Roy S, Flegal AR (2008) Mercury in the San Francisco estuary. Rev Environ Contam Toxicol 194:29–54

Conomos TE, Smith RE, Peterson DH, Hager SW, Schemel LE (1979) Processes affecting seasonal distributions of water properties in the San Francisco Bay estuarine system. In: Conomos TJ (ed) San Francisco Bay: The urbanized estuary. Reproduced from: Pacific Division AAAS, San Francisco, CA, pp 115–142. http://www.estuaryarchive.org/archive/conomos_1979. Accessed on 3 December 2009.

Connor M, Davis J, Leatherbarrow J, Werme C (2004) Legacy pesticides in San Francisco Bay: Conceptual model/impairment assessment. San Francisco Estuary Institute, Oakland, CA

Culberson SD, Foin TC, Collins JN (2004) The role of sedimentation in estuarine marsh development within the San Francisco estuary, California, USA. J Coast Res 20:970–979

Davis JA, Fry DM, Wilson BW (1997) Hepatic ethoxyresorufin-O-deethylase (EROD) activity and inducibility in wild populations of double-crested cormorants. Environ Toxicol Chem 16:1441–1449

Davis JA, May MD, Greenfield BK, Fairey R, Roberts C, Ichikawa G, Stoelting MS, Becker JS, Tjeerdema RS (2002) Contaminant concentrations in sport fish from San Francisco Bay, 1997. Mar Pollut Bull 44:1117–1129

Davis JA, Grenier JL, Melwani AR, Bezalel SN, Letteney EM, Zhang EJ, Odaya M (2007a) Bioaccumulation of pollutants in California waters: A review of historic data and assessment of impacts on fishing and aquatic life. Technical report of the Surface Water Ambient Monitoring Program, Oct 24, 2007. San Francisco Estuary Institute, Oakland, CA, 149 pp

Davis JA, Hetzel A, Oram JJ, McKee LJ (2007b) Polychlorinated biphenyls (PCBs) in San Francisco Bay. Environ Res 105:67–86

Dettinger MD, Cayan DR, Meyer M, Jeton AE (2004) Simulated hydrologic responses to climate variations and change in the Merced, Carson, and American River basins, Sierra Nevada, California, 1900–2099. Clim Change 62:283–317

Eagles-Smith CA, Ackerman JT, De La Cruz SEW, Takekawa JY (2009) Mercury bioaccumulation and risk to three waterbird foraging guilds is influenced by foraging ecology and breeding stage. Environ Poll 157:1993–2002

Evers DC, Savoy LJ, DeSorbo CR, Yates DE, Hanson W, Taylor KM, Siegel LS, Cooley JH, Bank MS, Major A, Munney K, Mower BF, Vogel HS, Schoch N, Pokras M, Goodale MW, Fair J (2008) Adverse effects from environmental mercury loads on breeding common loons. Ecotoxicol 17:69–81

Fairey R, Taberski K, Lamerdin S, Johnson E, Clark RP, Downing JW, Newman J, Petreas M (1997) Organochlorines and other environmental contaminants in muscle issues of sportfish collected from San Francisco Bay. Mar Pollut Bull 34:1058–1071

Federal Register (1970) Appendix D–United States list of endangered native fish and wildlife. Federal Regist 35:16047

Foxgrover AC, Higgins SA, Ingraca MK, Jaffe BE, Smith RE (2004) Deposition, erosion, and bathymetric change in South San Francisco Bay: 1858–1983. US. Geological Survey Open-File Report 2004–1192, 25 pp

Fuller CC, van Geen A, Baskaran M, Anima R (1999) Sediment chronology in San Francisco Bay, California, defined by ^{210}Pb, ^{234}Th, ^{137}Cs, and ^{239,240}Pu. Mar Chem 64:7–27

Gambrell RP (1994) Trace and toxic metals in wetlands: A review. J Environ Qual 23:883–891

Gilmour CC, Riedel GS, Ederington MC, Bell JT, Benoit JM, Gill GA, Stordal MC (1998) Methylmercury concentrations and production rates across a trophic gradient in the northern Everglades. Biogeochem 40: 327–345

Goals Project (1999) Baylands ecosystem habitat goals. A report of habitat recommendations prepared by the San Francisco Bay area wetlands ecosystem goals project. US. Environmental Protection Agency, San Francisco, CA; San Francisco Bay Regional Water Quality Control Board, Oakland, CA

Graczyk TK, Majewska AC, Kellogg JC (2008) The role of birds in dissemination of human waterborne enteropathogens. Trends Parasitol 24:55–59

Greenfield BK, Davis JA, Fairey R, Roberts C, Crane D, Ichikawa G (2005) Seasonal, interannual, and long-term variation in sport fish contamination, San Francisco Bay. Sci Tot Environ 336:25–43

Gunther AJ, Davis JA, Hardin DD, Gold J, Bell D, Crick JR, Scelfo GM, Sericano J, Stephenson M (1999) Long-term bioaccumulation monitoring with transplanted bivalves in the San Francisco Estuary. Mar Poll Bull 38:170–181

Hall BD; Aiken GR, Krabbenhoft DP, Marvin-DiPasquale M, Swarzenski CM (2008) Wetlands as principal zones of methylmercury production in southern Louisiana and the Gulf of Mexico region. Environ Pollut 154:124–134

Hammer DA, Bastian RK (1989) Wetland ecosystems: Natural water purifiers? In: Hamer DA (ed) (1989) Constructed wetlands for wastewater treatment: Municipal, industrial and agricultural. CRC, Boca Raton, FL, 831 pp

Heim WA, Coale KH, Stephenson M, Choe KY, Gill GA, Foe C (2007) Spatial and habitat-based variations in total and methyl mercury concentrations in surficial sediments in the San Francisco Bay-Delta. Environ Sci Technol 41:3501–3507

Heinz GH (1979) Methylmercury: reproductive and behavioral effects on three generations of mallard ducks. J Wildl Manage 43:94–401

Heinz GH, Hoffman DJ, Klimstra JD, Stebbins KR, Kondrad SL, Erwin CA (2009) Species differences in the sensitivity of avian embryos to methylmercury. Arch Environ Contam Toxicol 56:129–138

Hoffman DJ, Rattner BA, Bunck CM, Krynitsky A, Ohlendorf HM, Lowe RW (1986) Association between PCBs and lower embryonic weight in black-crowned night herons in San Francisco Bay. J Toxicol Environ Health 19:383–391

Holden A, She J, Tanner M, Lunder S, Sharp R, Hooper K (2003) PBDEs in the San Francisco Bay Area: measurement in fish. Organohalogen Compd 61:255–258

Hornberger MI, Luoma SN, van Geen A, Fuller C, Anima R (1999) Historical trends of metals in the sediments of San Francisco Bay, California. Mar Chem 64:39–55

Hothem RL, Roster DL, King KA, Keldsen TL, Marois KC, Wainwright SE (1995) Spatial and temporal trends of contaminants in eggs of wading birds from San Francisco Bay, California. Environ Toxicol Chem 14:1319–1331

Hunt JA, Davis JA, Greenfield BK, Melwani A, Fairey R, Sigala M, Crane DB, Regalado K, Bonnema A (2008) Contaminant concentrations in fish from San Francisco Bay. 2006. SFEI Contribution 554. San Francisco Estuary Institute, Oakland, CA

Hurley JP, Benoit JM, Babiarz CI, Shafer MM, Andren AW, Sullivan JR, Hammond R, Webb DA (1995) Influences of watershed characteristics on mercury levels in Wisconsin rivers. Environ Sci Technol 29:1867–1875

Hwang H-M, Green PG, Young TM (2009) Historical trend of trace metals in a sediment core from a contaminated tidal salt marsh in San Francisco Bay. Environ Geochem Health 31:421–430

Jaffe BE, Smith RE, Torresan LZ (1998) Sedimentation and bathymetric change in San Pablo Bay, 1856–1983. U.S. Geological Survey Open-File Report 98-759. U.S. Geological Survey, Menlo Park, CA

Jaffe BE, Foxgrover AC (2006) Sediment deposition and erosion in South San Francisco Bay, California from 1956 to 2005. U.S. Geological Survey Open-File Report 2006–1287, 24 pp. http://pubs.usgs.gov/of/2006/1287

Kelly CA, Rudd JWM, Bodaly RA, Roulet NP, St. Louis VL, Heyes A, Moore TR, Schiff S, Aravena R, Scott KJ, Dyck B, Harris R, Warner B, Edwards G (1997) Increases in fluxes of greenhouse gases and methyl mercury following flooding of an experimental reservoir. Environ Sci Technol 31:1334–1344

Kopec AD, Harvey JT (1995) Toxic pollutants, health indices, and population dynamics of harbor seals in San Francisco Bay, 1989–1992. Technical report of the San Francisco Bay Estuary Project. Moss Landing Marine Laboratories Tech Pub No 96-4, Moss Landing, CA, 172 pp

Krabbenhoft DP, Wiener JG, Brumbaugh WG, Olson ML, DeWild JF, Sabin TJ (1999) A national pilot study of mercury contamination of aquatic ecosystems along multiple gradients. In: Morganwalp DW, Buxton HT (eds) U.S. Geological Survey Toxic Substances Hydrology Program Proceedings of Technical Meeting, vol 2: Contamination of hydrologic systems and related ecosystems. U.S. Geological Survey Water Resource Investigations Report 99–4018B, pp 147–160

Levesque B, Brousseau P, Simard P, Dewailly E, Meisels M, Ramsay D, Joly J (1993) Impact of the ring-billed gull (*Larus delawarensis*) on the microbiological quality of recreational water. Appl Environ Microbiol 59:1228–1235

Levesque B, Brousseau P, Bernier F, Dewailly E, Joly J (2000) Study of the bacterial content of ring-billed gull droppings in relation to recreational water quality. Water Res 34: 1089–1096

Lonzarich DG, Smith JJ (1997) Water chemistry and community structure of saline and hypersaline salt evaporation ponds in San Francisco Bay, California. Calif Fish Game 83:89–104

Lowe S, Anderson B, Phillips B (2007) Final project report: Investigations of sources and effects of pyrethroid pesticides in watersheds of the San Francisco Bay Estuary. SFEI Contribution #523. San Francisco Estuary Institute, Oakland, CA

Manny BA, Johnson WC, Wetzel R (1994) Nutrient additions by waterfowl to lakes and reservoirs: Predicting their effects on productivity and water quality. Hydrobiologia 279:121–132

Marvin-DiPasquale M, Agee J, Bouse R, Jaffe B (2003) Microbial cycling of mercury in contaminated pelagic and wetland sediments of San Pablo Bay, California. Environ Geol 43:260–267

Marvin-DiPasquale M, Cox MH (2007) Legacy mercury in Alviso Slough, South San Francisco Bay, California: Concentration, speciation and mobility. U.S. Geological Survey, Open-File Report 2007–1240, 98 pp

McGourty CR, Hobbs JA, Bennett WA, Green PG, Hwang H-M, Ikemiyagi N, Lewis L, Cope JM (2009) Likely population-level effects of contaminants on a resident estuarine fish species: comparing *Gillichthys mirabilis* population static measurements and vital rates in San Francisco and Tomales Bays. Estuaries and Coasts 32:1111–1120

McKee LM, Ganju N, Schoellhamer D, Davis JA, Leatherbarrow J, Hoenicke R (2002) Estimates of suspended sediment flux entering San Francisco Bay from the Sacramento and San Joaquin delta. SFEI contribution 65. San Francisco Estuary Institute, Oakland, CA

McKee L, Oram J, Leatherbarrow J, Bonnema A, Heim W, Stephenson M (2006) Concentrations and loads of mercury in the lower Guadalupe River, San Jose, California: Water Years 2003, 2004, and 2005. SFEI Contribution 424. San Francisco Estuary Institute, Oakland, CA, 47 pp

Miles AK and Ricca MA (2010) Temporal and spatial distributions of sediment mercury at salt pond wetland restoration sites, San Francisco Bay, CA, USA. Sci Total Environ 408:1154–1165

Miller NL, Bashford KE, Strem E (2003) Potential impacts of climate change on California hydrology. J Am Water Resour Assoc 39:771–784

Neale JCC, Gulland FMD, Schmelzer KR, Harvey JT, Berg EA, Allen SG, Greig DJ, Grigg EK, Tjeerdema RS (2005) Contaminant loads and hematological correlates in the harbor seal (*Phoca vitulina*) of San Francisco Bay. J Toxicol Environ Health A 68:617–633

OEHHA (1994) Health advisory on catching and eating fish: Interim sport fish advisory for San Francisco Bay. Office of Environmental Health Hazard Assessment, California Environmental Protection Agency

Ohlendorf HM, Custer TW, Lowe RW, Rigney M, Cromartie E (1988) Organochlorines and mercury in eggs of coastal terns and herons in California, USA. Colonial Waterbirds 11:85–94

Oram JJ, McKee LJ, Werme CE, Connor MS, Oros DR, Grace R, Rodigari F (2008) A mass budget of polybrominated diphenyl ethers in San Francisco Bay, CA. Environ Int 34:1137–1147

Oros DR, Hoover D, Rodigari F, Crane D, Sericano J (2005) Levels and distribution of polybrominated diphenyl ethers in water, surface sediments, and bivalves from the San Francisco Estuary. Environ Sci Technol 39:33–41

Park J-S, Kalantzi OI, Kopec D, Petreas M (2009) Polychlorinated biphenyls (PCBs) and their hydroxylated metabolites (OH-PCBs) in livers of harbor seals (*Phoca vitulina*) from San Francisco Bay, California and Gulf of Maine. Mar Environ Res 67:129–135

Pereira WE, Hostettler FC, Luoma SN, van Geen A, Fuller CC, Anima RJ (1999) Sedimentary record of anthropogenic and biogenic polycyclic aromatic hydrocarbons in San Francisco Bay, California. Mar Chem 64:99–113

Reddy KR, D'Angelo EM (1997) Biogeochemical indicators to evaluate pollutant removal efficiency in constructed wetlands. Water Sci Technol 35:1–10

Reddy KR, Kadlec RH, Flaig E, Gale PM (1999) Phosphorus retention in streams and wetlands: A review. Crit Rev Environ Sci Technol 29:83–146

Sánchez-Chardi AM, López-Fuster MJ, Nadal J (2007) Bioaccumulation of lead, mercury, and cadmium in the greater white-toothed shrew, *Crocidura russula*, from the Ebro Delta (NE Spain): Sex-and age-dependent variation. Environ Pollut 145:7–14

Scherer NM, Gibbons HL, Stoops KB, Muller M (1995) Phosphorus loading of an urban lake by bird droppings. Lake Reserv Manage 11:317–327

Schwarzbach SE, Albertson J, Henderson J, Thomas C (2000) Influence of tidal marsh channel order on methylmercury bioaccumulation in California Clapper Rails of San Francisco Bay. Platform presentation, 21st annual meeting of the society of environmental toxicology and chemistry, 12–16 November, Nashville Tennessee

Schwarzbach SE, Henderson JD, Thomas SM, Albertson JD (2001) Organochlorine concentrations and eggshell thickness in failed eggs of the California clapper rail from South San Francisco Bay. Condor 103:620–624

Schwarzbach SE, Albertson JD, Thomas CM (2006) Effects of predation, flooding, and contamination on reproductive success of California clapper rails (*Rallus longirostris obsoletus*) in San Francisco Bay. Auk 123:45–60

SCVWD (2008) Draft environmental impact report. State Clearinghouse No. 2007082071. Alviso Slough Restoration Project. Santa Clara Valley Water District, May 2008

Selvendiran P, Driscoll CT, Bushey JT, Montesdeoca MR (2008) Wetland influence on mercury fate and transport in a temperate forested watershed. Environ Pollut 154:46–55

SFBRWQCB (2007) Total Maximum Daily Load for PCBs in San Francisco Bay: Proposed basin plan amendment and staff report. California Regional Water Quality Control Board, San Francisco Bay Region, Oakland, CA, December 2007

SFEI (2007) The pulse of the estuary: Monitoring and managing water quality in the San Francisco Estuary. SFEI Contribution 532. San Francisco Estuary Institute, Oakland, CA

SFEI (2009) The pulse of the Estuary: Monitoring and managing water quality in the San Francisco Estuary. SFEI Contribution 583. September 2009. San Francisco Estuary Institute, Oakland, CA

She J, Petreas M, Winkler J, Visita P, McKinney M, Kopec D (2002) PBDEs in the San Francisco Bay Area: Measurements in harbor seal blubber and human breast adipose tissue. Chemosphere 46:697–707

She J, Holden A, Tanner M, Sharp M, Adelsbach T, Hooper K (2004) Highest PBDE levels (max 63 ppm) yet found in biota measured in seabird eggs from San Francisco Bay. Organohalogen Compd 66:3939–3944

She J, Holden A, Adelsbach TA, Tanner M, Schwarzbach SE, Yee JL, Hopper J (2008) Concentrations and time trends of polybrominated diphenyl ethers (PBDEs) and polychlorinated biphenyls (PCBs) in aquatic egg bird eggs from San Francisco Bay, CA 2000–2003. Chemosphere 73:201–209

Shellenbarger GG, Athearn ND, Takekawa JY, Boehm AB (2008) Fecal indicator bacteria and Salmonella in ponds managed as bird habitat, San Francisco Bay, California, USA. Water Res 42:2921–2930

Sheoran AS, Sheoran V (2006) Heavy metal removal mechanism of acid mine drainage in wetlands: A critical review. Miner Eng 19:105–116

St. Louis VL, Rudd JWM, Kelly CA, Beaty KG, Flett RJ, Roulet NT (1996) Production and loss of methylmercury and loss of total mercury from boreal forest catchments containing different types of wetlands. Environ Sci Technol 30:2719–2729

Talmage SS, Walton BT (1993) Food chain transfer and potential renal toxicity of mercury to small mammals at a contaminated terrestrial field site. Ecotoxicol 2:243–256

Thomas MA, Conaway CH, Steding DJ, Marvin-DiPasquale M, Abusaba KE, Flegal AR (2002) Mercury contamination from historic mining in water and sediment, Guadalupe River and San Francisco Bay, California. Geochemistry 2:211–217

Thompson B, Adelsbach T, Brown C, Hunt J, Kuwabara J, Neale J, Ohlendorf H, Schwarzbach S, Spies R, Taberski K (2007) Biological effects of anthropogenic contaminants in the San Francisco Estuary. Environ Res 105:156–174

Thompson JK (2005) One estuary, one invasion, two responses – Phytoplankton and benthic community dynamics determine the effect of an estuarine invasive suspension-feeder. In: Dame RF, Olenin S (eds) The comparative roles of suspension-feeders in ecosystems. Springer, The Netherlands, pp 291–316

Tsao DC, Miles AK, Takekawa JY, Woo I (2009) Potential effects of mercury on threatened California Black Rails. Arch Environ Contam Toxicol 56:292–301

URS (2007) Status and trends of Delta-Suisun services. Prepared by URS Corporation for the California Department of Water Resources, March 2007, 54 pp

Valiela I, Alber M, LaMontagne M (1991) Fecal coliform loadings and stocks in Buttermilk Bay, Massachusetts, USA, and management implications. Environ Manage 15:659–674

Vanrheenen NT, Wood AW, Palmer RN, Lettenmaier DP (2004) Potential implications of PCM climate change scenarios for Sacramento-San Joaquin River Basin hydrology and water resources. Clim Change 62:257–281

van Geen A, Luoma SN (1999) The impact of human activities on sediments of San Francisco Bay, California: An overview. Mar Chem 64:1–6

Venkatesan MI, de Leon RP, van Geen A, Luoma SN (1999) Chlorinated hydrocarbon pesticides and polychlorinated biphenyls in sediment cores from San Francisco Bay. Mar Chem 64:85–97

Waldron MC, Colman JA, Breault RF (2000) Distribution, hydrologic transport, and cycling of total mercury and methyl mercury in a contaminated river-reservoir-wetland system (Sudbury River, eastern Massachusetts). Can J Fish Aq Sci 57:1080–1091

Windham-Myers L, Marvin-Dipasquale M, Krabbenhoft DP, Agee JL, Cox MH, Heredia-Middleton P, Coates C, Kakouros E (2009) Experimental removal of wetland emergent vegetation leads to decreased methylmercury production in surface sediment. J Geophys Res 114:G00C05

Index

A

Accidents, Baltic Sea shipping, 106
Accidents, oil spill causes (diag.), 97
Accidents, shipping oil spills, 95 ff
Accidents worldwide, tanker oil spills (table), 98
Alaska, lead–zinc mine, 49
Alaskan lead exposure, adults & children, 58
Animal adaptation role, HSP (heat shock protein) production, 8
Animal exposure to lead, Red Dog Mine, 56
Antarctic fish, muted heat-stress response, 10
Antioxidant enzymes, resist oxidative stress, 5
Antioxidants, oxidative stress mitigation, 4
Antioxidants, pollution biomarkers, 5
Apoptosis, HSP effect, 14
Aquatic organisms, HSPs as biomarkers, 17
Aquatic pollutants, oxidative stress, 4
Aromatic pollutant degradation, ring-hydroxylating oxygenases (RHOs), 65 ff

B

Baltic Sea, illegal oil discharges (illus.), 109
Baltic Sea, oil spills, 107
Baltic Sea, shipping accidents & oil spills (table), 107
Baltic Sea, shipping accidents, 106
Baltic Sea, shipping accident types (table), 108
Batie classes, RHOs, 67
Batie classification, nature & deficiencies, 68
Batie classification scheme, RHOs (table), 69
Bioavailability of hydrocarbons, marine species, 106
Bioavailability testing, mining wastes, 34
Biodegradation of oil spills, marine environment, 103
Biological aspects, San Francisco Bay water quality, 130

Biomarker, HSPs in fish, 17
Biomarkers of pollution, antioxidants, 5
Biomarkers of stress, aquatic contaminants, 4
Biomonitors, fish, 2
Bird contamination, San Francisco Bay, 123
Birds & water quality, San Francisco Bay, 137
Brominated diphenyl ethers (BDE), San Francisco Estuary (illus.), 127

C

Cadmium exposure, humans, 51
Catalytic pockets, RHO active sites (illus.), 83–84
Chemical aspects, San Francisco Bay water quality, 130
Chemical contaminants, south San Francisco Bay, 120
Chemical pollutants, HSP production, 7
Chemical stressors, in fish, 3
Children, lead exposure, 56
Classification systems, RHOs, 66
Consumption, petroleum products (table), 96
Contaminant erosion, buried sediments, 131
Contaminant removal, phytoremediation, 87
Contaminant sequestration in marshes, San Francisco Bay, 137
Contaminants, iron-ore tailings, 32
Contaminated soil contact, Red Dog Mine, 54
Contamination & stress, biomarkers, 4
Contamination, San Francisco Bay, 124
Contamination, south San Francisco Bay, 120
Cross (or protection) tolerance, HSP, 11
Crude oil, constituents, 98

D

Degradation improvements, RHOs, 85
Degradation of pollutants, oxygenases, 65 ff
Developmental role in fish, HSPs, 10
Dioxin contamination, San Francisco Bay, 127

Dispersion of oil spills, marine environment, 102
Dust-bearing soils, lead exposure (illus.), 55
Dust-borne metal exposure, Red Dog Mine, 53
Dust control, Red Dog Mine, 60
Dust exposure, Red Dog Mine, 49 ff
Dust and metal regulation, Red Dog Mine, 60

E

Effluents from mining, characteristics, 30
Electron transfer, RHOs, 78
Emerging contaminants, San Francisco Bay, 129
Emulsification in water, oil spills, 101
Environmental biomarker, HSPs in fish, 17
Environmental biomonitors, fish, 2
Environmental exposure studies, Red Dog Mine, 57
Environmental fate, marine oil spills, 100
Environmental impact, mining wastes, 33
Environmental implications, ship oil spills, 95 ff
Environmental pollution, stressed fish, 1 ff
Environmental stressors, HSPs, 2
Environmental stressors, impact on fish, 3
Environmental stressor types, fish, 3
Environmental stress, reactive oxygen species, 1
Epidemiology studies, Red Dog Mine, 57
Evaporation loss, oil spills, 101

F

Fate in environment, marine oil spills, 100
Fate of oil spills, marine environment (diag.), 101
Fate of oils spills, weathering, 100
Fish & pollution, heat shock proteins, 1 ff
Fish, antioxidant biomarkers of pollution, 5
Fish, chemical stressors, 3
Fish, developmental role of HSP, 10
Fish effects, spilled petroleum, 105
Fish, as environmental biomonitors, 2
Fish, environmental stressors, 3
Fish genes, HSPs, 11
Fish HSP expression, seasonal influence, 15
Fish HSP overexpression, stress defense, 17
Fish, impact of environmental stressors, 3
Fish, metal-induced stress, 4
Fish, oil spill effects, 104
Fish response, stress, 5
Fish, role of stress proteins, 8
Fish, stress hormone release, 6
Fugitive dust exposure, Red Dog Mine, 49 ff., 51

Future changes San Francisco Bay, water quality, 138

G

Generalized stress response, fish, 5
Geography, Red Dog Mine (illus.), 50
Glutathion, HSF (heat shock factor) relationship, 13

H

Heat shock factors (HSF), heat shock response, 12
Heat shock proteins (HSP), stressed fish, 1 ff
Heat stress proteins, stress response, 2
Heat-stress response, muted in Antarctic fish, 10
Heavy-metal human exposure, Red Dog Mine, 49 ff
Heavy metals, toxicity & exposure, 51
HSF, glutathion relationship, 13
HSF link, thiol-containing molecules, 13
HSP, apoptosis effect, 14
HSP, cross tolerance in fish, 11
HSP, developmental role in fish, 10
HSP expression, seasonal influences, 15
HSP families, molecular chaperones, 6
HSP genes, fish, 11
HSP (heat shock protein), description, 2
HSP induction, mechanistic regulation, 12
HSP overexpression in fish, stress defense, 17
HSP, pollutant protective response, 10
HSP production, chemical pollutants, 7
HSP production, role in animal adaptation, 8
HSP, protein metabolism interaction, 7
HSP, relationship to P450 inducers, 8
HSP, role in survival, 14
HSP, stress-induced production, 6
Human adverse effects, lead exposure (diag.), 52
Human exposure, contaminated soil contact, 54
Human exposure to metals, Red Dog Mine, 49 ff
Human exposure, zinc & cadmium, 51
Hydrocarbons from petroleum, water solubility, 99
Hyperaccumulator plants, listing (table), 39
Hyperaccumulator plants, for phytoremediation, 35
Hyperaccumulators, plant species, 42

I

Illegal oil discharges, Baltic Sea (illus.), 109
India, iron-ore beneficiation process, 32
India mining slimes, iron content (table), 33

Inducers of P450, HSP implications, 7
Inhalation of fugitive dust, Red Dog Mine, 53
International incidence, oil spills, 96
Iron content, Iron-ore slimes from mining (table), 33
Iron-ore beneficiation, India, 32
Iron-ore mining, metal toxicity, 31
Iron-ore mining wastes, phytoremediation treatment, 31
Iron-ore slimes, composition (table), 33
Iron-ore tailings, characterization, 32
Iron-ore tailings, composition, 32
Iron-ore tailings, contaminants, 32
Iron-ore tailings, nature and production, 31
Iron-ore tailings phytoaccumulation, lemon grass (illus.), 35
Iron-ore tailings treatment, tomato plants (illus.), 36
Iron-ore tailings treatment, tree species (illus.), 36
Iron-ore tailings utilization, phytoremediation, 34
Iron-ore wastes, from mining, 30
Iron-ore wastes, phytoremediation, 29 ff
Issues for San Francisco Bay, South Bay Salt Pond Restoration, 115 ff

K
Kweon classification scheme, RHOs, 70
Kweon classification scheme, RHOs (table), 71–72

L
Lead, in dust-bearing soils (illus.), 55
Lead exposure, children, 56
Lead exposure, Red Dog Mine, 52
Lead exposure, Red Dog Mine workers, 58
Lead exposure, subsistence land use, 56
Lead exposure, wildlife effects, 56
Lead health effects, humans (diag.), 52
Lead levels, Red Dog Mine workers (illus.), 59
Lead–zinc mine, Alaska, 49
Lemon grass, phytoremediation (illus.), 36

M
Mammals of San Francisco Bay, methylmercury contamination, 124
Marine environment, biodegradation of oil spills, 103
Marine environment, dispersion of oil spills, 102
Marine environment, fate of oil spills (diag.), 101
Marine environment, oil pollution, 97
Marine environment, oil spill effects, 97
Marine environment, sediment-oil interaction, 103
Marine oil spills, fate in environment, 100
Marine organisms, petroleum hydrocarbon accumulation, 106
Marine species, bioavailability of hydrocarbons, 106
Mercury contamination in fish, San Francisco Bay (diag.), 122
Mercury contamination, south San Francisco Bay, 121
Mercury residues, south San Francisco Bay sediment cores (diag.), 132
Mesocosms, managing mining wastes, 34
Metal contaminants, phytovolatilization clean-up, 38
Metal contaminants, subject to phytostabilization (table), 41
Metal content, *B juncea* plant uptake (table), 36
Metal exposure of humans & wildlife, Red Dog Mine, 51
Metal toxicity, fish stress, 4
Metal toxicity, from iron-ore mining, 31
Metal uptake, *B. juncea* (table), 36
Methylmercury, bird contamination, 123
Methylmercury contamination, San Francisco Bay, 123
Methylmercury, mammal contamination, 124
Methylmercury residue patterns, San Francisco Bay, 136
Methylmercury residue patterns, sediments (diag.), 135
Microbial degradation, role of ring-hydroxylating oxygenases, 66
Mineral oil effects, marine sea birds, 105
Mining slimes, composition (table), 33
Mining slimes, iron content (table), 33
Mining wastes, characteristics, 30
Mining wastes, environmental impact, 33
Mining wastes, minimization, 33
Mining wastes, remediation approaches, 35
Mining wastes, role for mesocosm studies, 34
Mining wastes, surface runoff, 30
Molecular chaperones, HSP families, 6
Mousse formation, oil spills, 101

N
Nam classification scheme, RHOs, 68
Nam classification scheme, RHOs (table), 70
Naphthalene dioxygenases (NDO), shunt reaction cycles (diag.), 80

O

Occupational exposure studies, Red Dog Mine, 58
Oil discharges, Baltic Sea (illus.), 109
Oil pollution, marine environment, 97
Oil spill biodegradation, marine environment, 103
Oil spill dispersion, marine environment, 102
Oil spill effects, dependent factors, 104
Oil spill effects, marine environment, 97
Oil spill effects, sea-birds, -fish & -animals, 104
Oil spill fate, marine environment (diag.), 101
Oil spills, Baltic Sea (table), 107
Oil spills, causes (diag.), 97
Oil spills, emulsification in water, 101
Oil spills, evaporation loss, 101
Oil spills, fate in marine environment, 100
Oil spills, implications, 109
Oil spills, international incidence, 96
Oil spills, major incidents, 97
Oil spills, mousse formation, 101
Oil spills, shipping accident implications, 95 ff
Oil spills, from tanker accidents (table), 98
Oil spill weathering, photo-oxidation, 102
Oxidative stress, antioxidant enzyme relief, 5
Oxidative stress, antioxidant mitigation, 4
Oxidative stress, aquatic pollution, 4
Oxygenases, aromatic pollutant degradation, 65 ff
Oxygenases, Rieske-type proteins, 67

P

P450 inducers, HSP implications, 8
PBDE (poly brominated diphenyl ethers) contamination, San Francisco Bay, 126
PCB (polychlorinated benzenes) contamination, San Francisco Bay, 124
Pesticide contaminants, San Francisco Bay, 128
Petroleum constituents, crude oil, 98
Petroleum hydrocarbon accumulation, marine organisms, 106
Petroleum products, sea water pollution, 95
Petroleum products, total consumption (table), 96
Petroleum transport, ship tankers, 96
Photo-oxidation, oil-spill weathering, 102
Physiological adaptations of stressed fish, HSPs, 1 ff
Phytoaccumulation of metals, *B. juncea* (table), 36
Phytoaccumulator plants, for phytoremediation, 35
Phytoextraction, phytoremediation approach, 37
Phytoremediation approach, phytoextraction, 37
Phytoremediation, characterization, 35
Phytoremediation, clean mining wastes, 31
Phytoremediation, contaminant removal, 87
Phytoremediation, iron-ore wastes, 29 ff
Phytoremediation, lemon grass (illus.), 35
Phytoremediation process, phytostabilization, 40
Phytoremediation process, phytovolatilization, 38
Phytoremediation process, rhizofiltration, 39
Phytoremediation, suitable plant species, 42
Phytoremediation, sustainable remediation approach, 34
Phytoremediation, tomato plants (illus.), 36
Phytoremediation, tree species (illus.), 36
Phytostabilization, candidate plant species (table), 41
Phytostabilization, mechanism depicted (illus.), 40
Phytostabilization, phytoremediation process, 40
Phytovolatilization, phytoremediation process, 38
Plant phytoextraction, phytoremediation approach, 37
Plants, hyperaccumulators of mining waste (table), 39
Plant species candidates, phytostabilization (table), 41
Plant species, capable of phytoremediation, 35
Plant species, hyperaccumulators, 42
Plant species, suitable for phytoremediation, 42
Pollutant degradation, oxygenases, 65 ff
Pollutant protective response, HSP production, 10
Pollution biomarkers, antioxidants, 5
Pollution, oil spill causes (diag.), 97
Polycyclic aromatic hydrocarbons (PAHs), sediment concentrations from oil (table), 104
Proteins, metabolism and HSPs, 6
Proteins, role in stressed fish, 7

R

Reaction cycles, NDO shunts (diag.), 80
Reactive oxygen species, environmental stress, 1

Reactive oxygen species (ROS), biological stress, 4
Red Dog Mine, contaminated soil contact, 54
Red Dog Mine, dust-borne metal exposure, 53
Red Dog Mine, dust control, 60
Red Dog Mine employees, blood lead levels (illus.), 59
Red Dog Mine, environmental exposure studies, 57
Red Dog Mine, epidemiology studies, 57
Red Dog Mine, heavy metal exposure, 51
Red Dog Mine heavy-metal exposure, Alaska, 49
Red Dog Mine, heavy-metal human exposure, 49 ff
Red Dog Mine, heavy metal toxicity, 51
Red Dog Mine, inhalation of fugitive dust, 53
Red Dog Mine, lead exposure, 52
Red Dog Mine, lead exposure effect, 56
Red Dog Mine, maps (illus.), 50
Red Dog Mine, occupational exposure studies, 58
Red Dog Mine, regulatory oversight, 60
Red Dog Mine, resident mine dust exposure, 55
Red Dog Mine, wildlife & human metal exposure, 51
Regioselectivity, RHOs, 81
Remediation approaches, mining wastes, 35
Resident exposure, mine dust, 55
Rhizofiltration, phytoremediation process, 39
RHO classification, Batie scheme (table), 69
RHO degradation techniques, improvements, 85
RHO pockets, structural residues (table), 82
RHO (ring-hydroxylating oxygenase) active sites, catalytic pockets (illus.), 83–84
RHOs, aromatic pollutant degradation, 65 ff
RHOs, Batie classes, 67
RHOs, classification systems, 66
RHOs, description, 66
RHOs, electron transfer, 78
RHOs, Kweon classification scheme, 70
RHOs, Kweon classification scheme (table), 71–72
RHOs, microbial degradation role, 66
RHOs, Nam classification, 68
RHOs, Nam classification scheme (table), 70
RHOs, regio- & stereo-selectivity, 81
RHOs, structural investigations, 72
RHOs, substrate oxidation, 78
RHOs, typical structures (illus.), 74
RHOs, α subunit structure, 75–78
RHOs, β subunit structure, 73
Rieske proteins, characterization, 67
Rieske-type proteins, oxygenases, 67

S

Salt pond restoration, San Francisco Bay, 117
San Francisco Bay, bird contamination, 123
San Francisco Bay, birds & water quality, 137
San Francisco Bay, contaminant sequestration in marshes, 137
San Francisco Bay, dioxin contamination, 127
San Francisco Bay, emerging contaminants, 129
San Francisco Bay mammals, methylmercury contamination, 124
San Francisco Bay, mercury contamination, 121
San Francisco Bay, mercury residues in fish (diag.), 122
San Francisco Bay, methylmercury contamination, 123
San Francisco Bay, methylmercury residue patterns, 136
San Francisco Bay, PBDE (polybrominated diphenyl ether) contamination, 126
San Francisco Bay, PCB (polychlorinated biphenyl) contamination, 124
San Francisco Bay, pesticide contaminants, 128
San Francisco Bay, salt ponds and wetland restoration, 117
San Francisco Bay sediment cores, south Bay mercury residues (diag.), 132
San Francisco Bay, south Bay contamination, 120
San Francisco Bay, South Bay Salt Pond Restoration Project (illus.), 119
San Francisco Bay, south Bay sediment erosion, 131
San Francisco Bay, south Bay water quality, 115 ff
San Francisco Bay, south Bay water quality, 120
San Francisco Bay, tidal marsh habitats (illus.), 118
San Francisco Bay water quality, biological aspects, 130
San Francisco Bay water quality, chemical aspects, 130
San Francisco Bay, watershed map (illus.), 117
San Francisco Bay, wildlife residues, 124
San Francisco Estuary, BDE (brominated diphenyl ether) contamination (illus.), 127

San Francisco Estuary, description, 116
Sea animals, oil spill effects, 104
Sea bird effects, mineral oil, 105
Sea birds, oil spill effects, 104
Seasonal influences, HSP expression, 15
Sea water pollution, petroleum products, 95
Sediment concentrations, polycyclic aromatic hydrocarbons (table), 104
Sediment cores, south San Francisco Bay contamination pattern, 134
Sediment-oil interaction, marine environment, 103
Sediments, contaminant erosion at depth, 131
Sediments, methylmercury residue patterns (diag.), 135
Shipping accidents & oil spills, Baltic Sea (table), 107
Shipping accidents, Baltic Sea, 106
Shipping accident types, Baltic Sea (table), 108
Shipping oil spills, environmental effects, 95 ff
Shipping oil spills, implications, 109
Ship tanker accidents, oil spills (table), 98
Ship tankers, petroleum transport, 96
Soil contact, Red Dog Mine contaminants, 54
South Bay Salt Pond Restoration Project, San Francisco Bay (illus.), 119
South Bay Salt Pond Restoration, San Francisco water quality, 115 ff
Spilled petroleum, fish effects, 105
Stereoselectivity, RHOs, 81
Stress biomarkers, aquatic contamination, 4
Stress defense, fish HSP overexpression, 17
Stressed fish, polluted environments, 1 ff
Stress hormones, fish, 6
Stress-induced production, HSP, 6
Stressors, impact on fish, 3
Stressor types, fish, 3
Stress proteins, role in fish, 7
Stress, response in fish, 5
Stress response, heat shock proteins, 2
Stress, role for molecular chaperones, 6
Structural nature, RHOs, 72
Structural residues, RHO pockets (table), 82
Structures, typical RHOs (illus.), 74
Subsistence land use, lead exposure, 56
Substrate oxidation, RHOs, 78
α Subunit structure, catalytic pocket (illus.), 78
α Subunit structure, RHOs, 75–78

β Subunit structure, RHOs, 73
Surface runoff, mining wastes, 30
Survival role, HSP, 14
Sustainable remediation approach, phytoremediation, 34

T

Tailings from iron ore, generation process, 31
Tailing wastes, iron-ore mining, 30
Techniques, RHO degradation improvement, 85
Thiol-containing molecules, HSF link, 13
Tidal marsh habitats, south San Francisco Bay (illus.), 118
Tomato plants, iron-ore tailings treatment (illus.), 36
Toxicity, heavy metals, 51
Toxicity testing, mining wastes, 34
Tree species, iron-ore tailings uptake (illus.), 36

W

Waste description, iron-ore tailings, 31–32
Wastes from iron-ore mining, characteristics, 30
Wastes from mining, phytoremediation, 29 ff
Waste utilization from mining, phytoremediation, 34
Water quality & San Francisco Bay, future changes, 138
Water quality, south San Francisco Bay, 115 ff
Water quality, south San Francisco Bay, 120
Watershed map, San Francisco Bay (illus.), 117
Water solubility, hydrocarbons from petroleum, 99
Weathering, fate of oil spills, 100
Wetland restoration, San Francisco Bay, 117–118
Wildlife contamination, methyl mercury, 123
Wildlife effects, lead exposure, 56
Wildlife exposure to lead, Red Dog Mine, 56
Wildlife exposure, Red Dog Mine, 51
Wildlife residues, San Francisco Bay, 124

Z

Zinc & lead concentrates, Red Dog Mine, 50
Zinc exposure, humans, 51

Breinigsville, PA USA
15 September 2010
245104BV00013B/4/P